Published in the United States of America
by Monasteria Press LLC, San Francisco

ISBN 978-1-7372582-1-6

CONTENTS

Have you ever heard music that is so touching that it hurts?

Have you ever heard music that is so touching that it hurts? It is gently flowing into your head, slowly expanding in your heart, gaining momentum and amplifying your senses - until you feel beauty, love, happiness, magic - all at once. And as you close your eyes, you can see eternity.

We need an orchestra to infuse a greater force into a melody – not a single trumpet, not a single violin, and not 10 violins. It takes an orchestra to play a symphony.

Likewise, we need diverse players in medicine and biomedical research. Historically, the field of science, technology, engineering, mathematics and medicine (STEM) has been overwhelmingly male and overwhelmingly white. In the future, we will need more diverse teams to serve patients in a multi-cultural environment, to provide diverse role models for our team members, to unfold discoveries at the interface of different disciplines and to thrive in a constantly changing multi-faceted world.

Lu Hong and Scott Page showed that diverse groups of problem solvers outperformed high performing individuals and uniform groups of high performing problem solvers. In health care, diverse backgrounds, experiences and communication skills are critical to serve our increasingly diverse patient population, to solve daily challenges and to cure humanity – one patient at a time.

Anyone who has tried to conduct an orchestra knows how hard it is to bring the ensemble together: The trumpets may be overbearing, the violins may not play in harmony, the flute might not get heard. Sounds familiar? In medicine and biomedical research, we want to unite diverse team members with a wide range of abilities to create symphonies of medical care and innovation. To do this, we have to start with a mutual understanding of appreciation and respect for every team member. This will enable us to unleash every team members full potential and combine our unique skills towards a collaborative performance that is larger than any one of us could have achieved alone.

This book has been written for health care professionals, scientists and administrators, who want to increase their cultural competency and become effective allies for trainees, colleagues and patients in the field of clinical medicine and biomedical research. Insights shared here will be also useful for students, staff and faculty who are working in other STEM fields. Readers of this book will develop an increased awareness and understanding of cultures and thought processes that are different than their own and thereby, increase their effectiveness in interpersonal communications and negotiations.

This book is composed of 11 chapters which provide detailed insights into different aspects of diversity, such as race/ethnicity, sex and sexuality, age, marital status, economic background and ability/disability, among others. The book also discusses strengths and needs of summer students and first generation college graduates, who are new to the academic environment. The book challenges the reader to empathetically understand - rather than dismiss - different ideological viewpoints. Lessons learned will culminate in a celebration of diversity of thought as the engine of empathy, collaboration and innovation.

Each chapter in this book starts with a summary of a specific diversity topic, followed by a list of actionable items that can help to realize diversity related opportunities and address topic-specific biases and barriers. These actionable items were derived by discussion groups on these topics and are meant to provide the reader with examples, how they can make a positive difference. It is well possible that suggestions for one aspect of diversity (e.g. women in STEM) might also be beneficial to another (e.g. racial equity). The ultimate goal is to inspire the reader to create their own action plan and foster diversity and inclusion in their own community and/or at their own institution.

Since no single person can comprehensively reflect the broad range of different viewpoints and experiences of different members in our community, I invited my team members at Stanford Medicine to share their unique thoughts, experiences and perspectives. Most of the contributors are based in the Department of Radiology, while some are based in other clinical Departments, basic science Departments and/or other Institutions. Many of the stories that these people generously shared reveal that we are much more alike than we are different. Other stories and commentaries may provide insights into experiences and beliefs that are different from those of the reader. When reading these stories, I would like to encourage the reader to suspended judgment and try to develop an open mind for different points of view. It does not mean that one cannot have views of their own. The goal is rather to develop the ability to consider the merit of what other people communicate before drawing any conclusions. This ability to listen first, integrate the provided information into existing knowledge, and then drawing conclusions will be a most impactful skill that readers can learn from interactions with this book.

As our authors explain, some members of minority groups have experienced severe offenses by majority groups. Oppression in the environment of academic medicine can be associated with economic exploitation and lack of power of the minority group, unearned (not earned) advantages of the majority group, and biased thought processes and actions that discriminate and deprive the minority group. If minority members have to prove their worth over and over again, they cannot focus their creative energy towards productive contributions. This book aims to inspire a collective understanding of shared values in order to create a culture of fulfillment, creativity and prosperity for everyone.

To unite diverse minds to inspire symphonies of medical innovation, our book builds upon shared values of inclusiveness and equal opportunity. Diverse opinions and ideas are essential to the progress, growth and prosperity of medicine and its microcosmoses. The stories in this book should help the reader to recognize unique skills, talents and insights of people from diverse backgrounds such that we can broaden our collective perspectives, collaborate effectively and create a better future for all of us.

H. E. Daldrup-Link

Heike E. Daldrup-Link, MD
Professor and Associate Chair for Diversity
Stanford Medicine | Radiology

"When we listen and celebrate what is both common and different, we become wiser, more inclusive, and better as an organization.."

– Pat Wadors

Who We Are and What We Stand For

The faculty, staff, and trainees of Stanford Radiology reaffirm the Department's commitment to fostering a diverse and welcoming learning community that wholeheartedly embraces all of its members, including those who have been marginalized by aspects of their identities. We believe that a broad definition of diversity is critical for the protection of human rights and human dignity. The principles of respect and inclusion are essential to our joint mission, as an academic institution, that is dedicated to both cultivating dialogue across differences and to the development of meaningful citizenship that will meet the challenges of a highly complex society. To achieve our goal of creating a more diverse Department of Radiology community, we rely on, and therefore foster, diversity in the faculty, the staff, and our trainees. We value diversity in areas such as race, color, national or ethnic origin, sex, sexual orientation, gender identity, religion, age, parental or marital status, veteran status, and disability. Therefore, the department places a high value on recruiting and retaining individuals who can contribute to the department and Stanford University diversity goals, in all hiring. We are also committed to continual learning and improving and always strive to do better. When we are working on diversity, it should not be part of some to-do list. It is a reality that should be deeply felt and valued by all of us and eventually become engrained in our very cultural fabric.

Diversity can potentially be a sensitive topic. A healthy conversation on diversity starts with giving voice to individuals from a variety of perspectives and backgrounds. The experiences they share often reveal both strengths and imperfections in people and in organizations. This book, borne from the newsletter of the Department's Diversity Committee, chaired by Dr. Heike Daldrup-Link, contains many examples of courage and inspiration, revealing some opportunities for improvement, but mostly providing reflections that are simply meant to be appreciated on their own merits. This collection reflects unfiltered perspectives of a broad array of individuals. While no single individual's perspective represents that of the entire Department, the combination of diverse narratives helps establish the candid dialogue necessary to build a culture of inclusion and support.

Sam Gambhir

Sanjiv Sam Gambhir, MD, PhD
Professor and Chair
Stanford Medicine | Radiology

Notes

UNDERREPRESENTED MINORITIES IN MEDICINE

"Diversity is not about how we differ.
Diversity is about embracing one another's uniqueness."

– Ola Joseph

Our first chapter is dedicated to faculty, trainees and staff from racial and ethnic underrepresented minority backgrounds. We celebrate and appreciate their dedication, hard work and important contributions to the field of medicine and science, technology, engineering and mathematics (STEM). We also present data and personal reflections, which show that we have more to do.

Privilege is blind. People who have been protected from marginalizing experiences of minority groups might not understand the emotional strength, resilience, and energy needed to overcome stereotypes and cultural barriers. This chapter shows how we can support each other and create a more inclusive environment that will benefit us all. Martin Luther King said: "stop judging people by the color of their skin and start judging them by the content of their character."

Throughout this chapter and the entire book, we are presenting notable quotes to the interested reader in order to distill clarity from lived experiences, stimulate the mind, invite personal reflections and inspire change. When you reach a quote that you connect with, take a pause, reflect, acknowledge your own thoughts and derive wisdom for your specific situation and your unique community.

It can be humbling to realize the enormous potential of a bold, new perspective. What did most African Americans do decades ago if they were told to go to the back of the bus? Rosa Parks decided: The answer is no. And she changed a whole society.

How to be an ally

Many of us are struggling with what we can do as individuals in a political, biased environment, where racism and xenophobia remain perversive. Here, we would like to share a few actionable items that worked for our community. These points provide examples, are not meant to be comprehensive and may require adjustments for different communities and different contexts:

- Educate yourself and your community through lectures, newsletters, journal clubs etc.
- Understand the difference between cultural humility and cultural competence
- Recognize bias and discrimination against Black Americans and other people from racial/ethic minority backgrounds. Speak up if you notice intentional or unintentional microaggressions. Identify power imbalances and advocate for others.
- Provide members from underrepresented minority backgrounds with opportunities to share their experiences and ideas. Listen.
- Leaders on "listening tours": Ask if people who share their ideas would like to stay anonymous or if they would like to be referenced. Do not steel ideas.
- Increase the representation of minority members in your Department / institution through pipeline programs, outreach efforts and recruitment efforts
- Provide paid internship opportunities for students
- Create pathways for new growth and leadership.
- Invite speakers from diverse backgrounds for Grand Rounds lectures
- Increase the representation of minority members in leadership positions
- Monitor diversity metrics and potential disparities with regards to access to resources, work assignments, leadership opportunities, career advancements, salaries
- Create an environment that supports an inclusive culture
- Support networking events, actively connect team members
- Establish peer mentorship programs: Match mentees with mentors who are just one step ahead of them in their career
- Facilitate sponsorship by senior leaders by establishing regular 1:1 meetings between mentees and established leaders in the department /organization
- Organize imposter syndrome workshops
- Establish an office and/or "go to" person for confidential consultations regarding diversity matters and mental health matters
- Introduce a process to address concerns of bias and microaggression, e.g. a peer learning conference where anonymous concerns about interpersonal interactions are discussed
- Organize bidirectional town hall meetings and brainstorming sessions to seek feedback how diversity and inclusion in your specific community can be advanced
- Introduce mandatory regulatory anti-racist and anti-bias training
- Conduct regular climate surveys
- Measure and address racial and ethnic health disparities. Provide incentives for high quality medical care, impactful innovations and improved health outcomes
- Fund seed grant programs for diversity related projects
- Recognize time spent on diversity initiatives in career advancement deliberations
- Publicly value and appreciate team members for their contributions and achievements
- Create a sense of hope and optimism for the future; inspired by the openness and sense of justice of younger generations.

Generalizations and Stereotypes

Thank you for collecting input from black members of the Department. I would first like to say that many of my non-black colleagues are very supportive. However, it is also important to note that black people in STEM are still facing racial bias, stereotypes and microaggressions on a daily basis.

Examples: I am regularly asked to present my ID for verification by security personnel or even strangers. When I go into the book shop, the security person "discretely" followed me on multiple occasions. When I travel, I plan for extra time at the airport because I am regularly the subject of "random screenings". All these actions tell me very clearly: You do not belong here.

I receive embarrassing praise when I complete simple academic tasks. As if this were a surprise. Staff, students and faculty feel entitled to question my comments or actions, to correct my speech or my work on a regular basis, and to provide unsolicited advice about how I could do better. My role is regularly questioned by support personnel. For every simple task that I need assistance with, I have to explain why and when I need it. On multiple occasions, a staff person who just completed the same task for my white colleague tells me that they cannot complete it for me for some dubious reason.

At meetings, I am often consulted about diversity matters. I am automatically assumed to be the expert on diversity even though I have no specific training in human resource matters. When a black colleague has a problem, I am immediately consulted as an expert on the matter. He/she is also black, so obviously I should know what they were thinking. I don't recall how many times I have been asked about my dancing or athletic abilities. I do not have them.

Thanks for asking. It is important to learn from each other. Next, we need to do something about it.

Anonymous

"Growing up, I decided, a long time ago, I wouldn't accept any man-made differences between human beings, differences made at somebody else's insistence or someone else's whim or convenience."

– Maya Angelou

> *Change will not come if we wait for some other person or some other time. We are the ones we've been waiting for. We are the change that we seek.*
>
> — Barack Obama

Challenges faced by Black/African-American people in academia

Maxine Umeh Garcia, PhD
Postdoctoral Research Fellow

Below are my take on three challenges faced by Black/African-American people in academia. I am sure there are more, but these stood out to me. I wrote pretty freely/openly and from my own experiences. I feel that it is hard to identify specific problems without using the blanket "racial bias/racism" term, but hopefully sharing some of the challenges I have faced as a woman of color in STEM will help identify larger issues.

SEVERE IMPOSTER SYNDROME

On some level everyone deals with imposter syndrome, especially in academia and at top tier institutions like Stanford. But imagine being the only black person in almost every science class you've ever taken, being able to count on your fingers (on one hand) the number of black PhD students in your graduate group in the 6 years you had been in the program, or still, to this day, getting stared at, or people doing double-takes, when you walk the halls of your lab building especially if you are not wearing your University-issued badge. This is my reality. Over time you become somewhat desensitized, thankfully. But when I slow down long enough to think about my experiences I realize that I have been told verbally and nonverbally, to my face and behind my back, that I am an imposter. That I do not belong here. And given the number of diversity efforts that have still resulted in clear underrepresentation of Black full professors in STEM, it tells me that I will never really belong here. It makes you question yourself, all the way to your core. It is hard to process and be at peace with the fact that this part of me, that I cannot (and would not) change, could affect my ability to get a second interview, or receive R01 funding, or gain tenure. That it could negatively affect my ability to reach my career goals.

THE DOUBLE-EDGED SWORD/'KNOW YOUR PLACE' MICROAGGRESSIONS

As a senior PhD student I sat in a room with a handful of other underrepresented minorities (URMs) in our graduate program to discuss possible reasons why all of the URMs in our 1st year cohort were failing their two core courses. As I walked away from the meeting and continued to process, I later emailed my graduate group coordinator and shared what I felt was a summation of my experiences: "If I succeed it's because I am a URM. If I fail it's because I am a URM."

If I succeed in becoming an independent researcher in STEM it is only because I was afforded opportunities or given "special treatment" through diversity fellowships and programs, not because of my intellect, work ethic, and merit. But if I fail to reach my career goal it is because I am a minority and I was ill-prepared to keep up with the rigor of my PhD program or my skills were less than my counterparts. So you see, I stand in a lose-lose situation. Irrespective of the end results, whether positive or negative, the outcome is attributed to my "Blackness." This attribution is what I have heard termed "know your place racism/microaggressions". This is the concept that as a Black person, whether I succeed or fail I am reminded/told that it is due to my Blackness. As such, we are continually confronted with the reminder that we are Black. That being Black determines my actions. That being Black determines my outcomes. That being Black determines my success.

and failures. That being Black determines what I can afford and what I am afforded. That being Black means that I am less than others. And when I dare rise above my Blackness, to achieve, to attain something that has nothing to do with my Blackness, I am again reminded that I should 'know my place', because of course my success only came about because I was given special treatment because I am Black.

A student in my PhD cohort actually said to other students, "you know they only got the spots because they are minorities." (Referring to the few trainee positions on a T32 fellowship for which many students had interviewed.) It was not because I had practiced my chalk talk over and over and over until it was flawless, or that the interviewing committee never stopped me to ask clarifying questions and expressed how impressive my talk had been. What only mattered was my Blackness, my minority status, my "usefulness" for meeting a diversity quota. And trust me when I say that it begins to eat away at you, makes you question yourself. Maybe I am an imposter? May I have only gotten this far because I am Black? Not because I am hard working, or intelligent, driven, ambitious, creative, or a great communicator. I spent many hours in my PhD advisor's office questioning whether I could even make it to a place like Stanford because "maybe without these diversity initiatives I would have never stood a chance." I have since had to fight to overcome those doubts and believe that I am here because of my merit and not just my skin color.

SHOULDERING AN INVISIBLE BURDEN

Due to the numbers, or lack there of, of Black grad students, postdocs, and professors, at some point in your academic career you have probably been the "token" Black person on a committee, panel, or program event. I know that I have, many times. While there is nothing wrong with that. (I would venture to say that many URMs, including myself, enjoy participating in these things, getting the opportunity to share our experiences and educate others). However, because of this, we often represent, or it is insinuated that we represent, all Black people in academia. Which, again, is not necessarily a bad thing. But it means that I now carry the burden of behaving, performing, and excelling in such a way that I positively represent not only Blacks in STEM/academia, but the entire Black community. Because the truth of the matter is that I may be the only Black grad student, post-doc, and/or professor that a person meets or interacts with. Therefore, whether intentional or not, I carry that weight with me throughout my academic career. The risks of my failures are applied to the whole group. However the benefits of my success are only applied to me. And yet I continue to shoulder this burden as a means of proving not only my worth as a grad student, postdoc, and/or professor, but the worth of the black community. This means that I find myself working twice as hard as my counterparts just so I can be seen as equal to them. It means I never risk slacking off, rarely say no to service, and am constantly/consistently checking over my shoulder and to my sides to make sure I am running at the same pace as my counterparts, even though I am carrying extra weight. Always having to prove that I am here because of my merit and not just because I am the token minority. Truthfully, it is exhausting, and I do my best to remind myself that I don't have to carry this invisible burden. But the truth is until we see better representation of Blacks in STEM, the burden will remain.

Maxine Chidinma Umeh Garcia, MSc, PhD

Postdoctoral Research Fellow
Stanford Medicine | Neurosurgery

Growing up, I was blind to my own White Privilege. I had the naïve notion that my elementary school friend, Kelery, and I were equal. I thought that slavery, segregation, and discrimination were a part of history – that they lived only within the past. I believed that if I saw the color of a person's skin, then I was racist. I falsely thought that there was liberty and justice for all – that we were all equal. Shamefully, at one time, I did not understand the need for affirmative action. And even more shamefully, even though I was not against it, I did not know how desperately our Nation (and our world for that matter) needed a Black Lives Matter movement. Moreover, I did not know or even understand how an entire race could fear the police. The most I feared when being pulled over was that I was going to get a ticket for speeding – when I was in fact doing just that, speeding. In the words of Peggy McIntosh: "In my class and place, I did not see myself as a racist because I was taught to recognize racism only in individual acts of meanness by members of my group, never in invisible systems conferring unsought racial dominance on my group from birth". (1) I used to think that the White people who had been slave owners or who had pillaged or killed members of the Black community were the ones who had to pay. However, I now see that even though I never asked for it, I have unknowingly taken from and oppressed the Black community by the mere nature of my skin. I owe them and want to do my part to dismantle the social contract and give back to the Black community what is rightfully theirs – real equal opportunity and pursuit of life, liberty, ownership, and justice.

Over the years, I have learned that affirming that we are not the same and that I see color means that I see the oppression, the mistreatment, the pain, and the suffering of the Black community, the Latino community, and all communities who are dominated on the basis of the color of their skin or where they come from. It means that I understand the inalienable need for the Black Lives Matter movement. It means that even though I may never be able to truly understand the suffering of those oppressed, I know that I must be part of the Black Lives Matter fight. I must not be silent. Yes, we are all humans. Yes, we all deserve equal rights. But until we all acknowledge our differences, including the privilege of one race over another, then we cannot deconstruct the broken social systems that continue to propagate injustice. We must acknowledge that a racial contract exists. And like Adam Serwer so perfectly stated, we must acknowledge that "the racial contract is not partisan – it guides staunch conservatives and sensitive liberals alike – but it works most effectively when it remains imperceptible to its beneficiaries. As long as it is invisible, members of society can proceed as though the provisions of the social contract apply equally to everyone." (2) And prior to learning the truth, I unfortunately believed that the social contract applied equally to everyone. "But when an injustice pushes the racial contract in to the open, it forces people to choose whether to embrace, contest, or deny its existence." (2) And yes, we can and must acknowledge the exis-tence of the racial contract and then dismantle it at its core to build a more just world.

Kristina Michaudet, MD

Resident
Stanford Medicine | Radiology

References

1. *McIntosh, P. White Privilege: Unpacking the Invisible Knapsack. Peace and Freedom Magazine. 1989; July/August: 10-12.*
2. *Serwer, A. The Coronavirus Was an Emergency Until Trump Found Out Who Was Dying. The Atlantic. 2020; May, 8. https://www.theatlantic.com/ideas/archive/2020/05/americas-racial-contract-showing/611389/*

New Concepts About Racial Equity

Through the Department of Radiology's Racial Justice Challenge, I've been introduced to new concepts about racial equity, or the lack thereof. My K-12 education was "old" enough that I was taught the government always has the people's best interests at heart, and if you follow rules and laws, that you will be fine. In addition, I had heard about the concept of reparations, but didn't understand why history had to be relived. We no longer lived in a time of slavery or segregation. Why couldn't we just go forward in our improved and fair society? A few weeks ago, I read the article, "The Case for Reparations" by Ta-Nehisi Coates. It opened my eyes on multiple levels. First, it taught me that the government isn't always fair. In fact, it is sadly often the perpetrator in systemic racism. After all, a government is the combined voices of a group of people, and if that group of people is of a mind to oppress another group, it may be done under a facade of legality. It also helped me understand the impact that fewer resources can have on future generations. How, for Blacks, those resources have been diminished repeatedly, from the theft of Black-owned land in the early part of the 1900s, to the impact of redlining and the resulting predatory loan programs that sprang up in the middle of the same century. The inability to build a solid foundation for future generations to rely on has led to more uncertain footing for Blacks than other groups, and the repercussions of that has lasted for generations. And finally, I've begun to understand that the concept of reparations isn't necessarily or only about compensation for past losses, but a recognition of past mistakes and an advocacy for studying the issue. Through this Racial Justice Challenge, my education has become more rounded, and I'm more comfortable in acknowledging that my prior beliefs were founded on a skewed education system. I also recognize that some of my beliefs today may need to be revised in the future. I look forward to continuing my journey in this space.

Deepa Basava

Director, Finance and Administration
Cancer Center Faculty Practice
Stanford Medicine | Radiation Oncology

A Tale of Two African Journeys to the USA

Alida and Benedict were born in Cameroon, a bilingual French and English speaking Central African country nick-named "Africa in miniature" because it is richly blessed with a vast variety of geographical landscape, agricultural products, minerals and natural resources that mimic and mirror the continent. Cameroonians come from very diverse cultural backgrounds. Alida and Benedict both hail from the English-speaking part of the country which comprises only about 20% of the population. Theirs is a story of two people born into a minority and marginalized population, driven by the lack of opportunity to travel abroad against all the odds in search of better lives. Along the way, their emigration paths crossed. They met, fell in love and got married in the U.S. Their separate migration stories are characterized by uncertainty and luck that is guided by stubborn faith, sound mentoring and hard work.

After graduating with a bachelor's degree in Cameroon, Alida tried looking for a job for almost three years in vain. As the first daughter with five siblings, she wanted just to help support her family and never really thought of emi-grating abroad because it was a pipe dream due to her poor background. She had to do business, buying and selling food stuff to help support her family. One fine May afternoon in 2008, she received a call from a very good child-hood friend who had played the U.S. Diversity Visa Lottery on her behalf using an old passport size photo without even informing her. She had won the lottery and with this, her journey to becoming a US citizen, though controlled by unforeseen circumstances, became a reality. A minority-based program often criticized by many as being random turned out to provide opportunity and hope to a young graduate from a country currently in the pangs of serious political turmoil, resulting from the marginalization of one-fifth of its population regarding opportunities and re-sources over time. Alida is currently studying to become a nurse in the U.S.

Benedict's journey to Stanford was driven by a passion for math and science. He lost his mum to recurrent breast cancer in high school, and one of her last wishes was that "he should continue to be passionate about maths and science ..." After graduating with a BSc in math and due to lack of opportunities for further studies back home, he travelled abroad. In Belgium, he was accepted into an MSc degree program in biostatistics. As a foreign student, like in most countries, you are only allowed to work a limited number of hours per year. While this condition is meant to help foreign students focus on their studies, it is oblivious to the fact that many foreign minority students are expected to help support their families back home while studying. To raise more money to help his ailing dad and at the same time concentrate fully on his studies, he opted to pick fruit in farms on the outskirts of cities like Antwerpen during holidays in cold and harsh weather conditions. Some of his classmates were shocked at how much weight he had lost at the beginning of each academic year, a consequence of grueling and punishing schedules. Despite all, he kept his grades up. The good news after graduating from Belgium was that he got a scholarship for his Ph.D. studies in Germany. He was the only black individual in the program, and his German was terrible. His scientific mentor was very supportive, and this support made it possible for him to get fellowship awards to travel to the U.S. to present his research. During one of these visits, he met Alida and the rest is history. He later graduated with a Ph.D., then applied and was accepted to an NIH funded postdoctoral program at Stanford. He travelled to the U.S. as a postdoctoral scholar and is now an instructor in the Department of Radiology. He is happily married to Alida with two kids and is passionately pursuing a dream that started about 20 years ago with the words of a dying mother who saw a vision and a father who inspired.

The life of an underrepresented minority (URM), especially in a foreign country, is full of uncertainty and tough choices that often have to be made, usually with limited resources. While the outcome of these decisions can be very diverse with some good and some bad, there is something common triggering minorities to emigrate: the lack of opportunities and the pursuit of elusive happiness that is promised in the American Declaration of Independence.

Benedict Anchang, PhD

Instructor
Stanford Medicine | Radiology

"*. . . continue to be passionate about maths and science . . .*"

1,000 Inspiring Black Scientists in America

Maya Angelou once said, "We delight in the beauty of the butterfly, but rarely admit the changes it has gone through to achieve that beauty." This can be said for Black culture in its entirety.

Black culture is the archetype for innovation, distinction, creativity, intelligence, spirituality, and healing. Black culture is distinct, and it has a heavy influence on American and global cultures. Likewise, Black talent is both ubiquitous and abundant, but the excellence of Black people is often obscured. To remove the bleach from the history books, the Community of Scholars formed this list of inspiring Black scientists. We are also here to dismantle the myth that outstanding Black scientists make up a small percentage of the scientific community.

http://crosstalk.cell.com/blog/1000-inspiring-black-scientists-in-america

BEING THE CHANGE:
Meet Dar es Salaam

Dr. Jayne Seekins is a pediatric radiologist at Lucile Packard Children's Hospital, who lived in Dar es Salaam with her family for several years. Her husband was the Senior Defense Official and Defense Attaché at the Embassy of the United States of America in Dar es Salaam. Tanzania is the 5th African nation that Dr. Seekin's family has lived in. She had previously lived and volunteered in Cameroon, Mali, Burkina Faso, and Cote D'Ivoire.

Muhimbili National Hospital is a large multi-specialty hospital complex that sits just outside of downtown Dar es Salaam, Tanzania. It is comprised of many buildings over a large campus. There is the main hospital with the Emer-gency Department, the Cardiac Institute, Muhimbili Orthopedic Institute, a Children's Hospital, and a Maternity Hospital.

Included in the campus are the School of Medicine, Nursing, Midwifery, Medical Technology and many more. Currently, there are 43 Radiology residents who rotate through the Radiology department. They also rotate through the other three national hospitals in Dar. This is one of only two Radiology residency programs in the country, and it is the largest.

Muhimbili is truly the cornerstone of health care for the country. As this is the referral center for the nation, Tanzanians come from all corners of the country to Muhimbili for care. They may have traveled tens or hundreds of miles to their local health care center to then wait for the transfer to Dar es Salaam.

Dr. Jayne Seekins explains: *"In supporting the learning of the radiology residents and faculty, I do much more than helping with a complex diagnosis. We discuss workflows, quality improvement, report structuring, improving imaging protocols and so much more. I am fortunate enough to be there when in the country. The eagerness for collaboration is palpable. I am now trying to raise funds so that our Radiology residents, colleagues and staff could come to Tanzania and volunteer with me. I am sure that this experience would be invaluable for both parties."*

Tanzania Reflection

In October-November 2018, Justin Tse, a PGY-4 radiology resident at Stanford University, completed a 2.5 week rotation at Muhimbili National Hospital in Dar es Salaam in Tanzania. He joined a team of radiology residents and attendings from Dartmouth and Yale to help improve the practice of radiology in Tanzania. Justin will be finishing his residency at Stanford in 2020 and will be pursuing a fellowship in abdominal imaging and intervention afterwards. He hopes to combine his interests in abdominal radiology and global health by expanding HCC screening in low- and middle-income countries where hepatitis B is endemic, particularly sub-Saharan Africa. Here are his insights and reflections from his trip:

What were your first thoughts upon arriving in Tanzania?

If you can afford a flight to Tanzania, you will automatically be among wealthiest 5% in the country. That's the percentage of the country's population that has a bank account. The unemployment rate among young people was estimated to be 86% in 2012. And if you had a job, it'd pay about 86 cents an hour.

If you end up in Tanzania, you'd also be one of the oldest. The median age of the country is 17, and over 75% are estimated to be under the age of 25. A country of very young and unemployed (coupled with an all cash economy) is disastrous. Without income, there are no effective taxes; without taxes, there is no capital to fund social services, infrastructure, and government workers. When you have poorly paid government workers who are also assigned authority, it's fertile ground for bribery and corruption. This is what I had mentally prepared myself for prior to my arrival.

I landed in Dar es Salaam following a 2-day journey, first flying from SFO to Zurich (11 hours), Zurich to Nairobi (9 hours), and finally Nairobi to Dar es Salaam (1.5 hours). Despite a nighttime arrival, it was still at least 80 degrees and humid. This would normally be perfect t-shirt and shorts weather in CA. Nevertheless, I made sure to cover up everything except my face, both to respect local religious/social customs of modesty and as a defense against mosquitos (the causative species of cerebral malaria is most rampant at night).

The first thing I did when I landed was apply for a visa. This ended up being the most uncertain part of my trip. "Volunteer", as it turns out, was recently removed as a reason to enter the country. The reasoning I'm told is two-fold. First, the presence of volunteers suggests that the country is in need of aid, which the government hates to admit. Second, volunteer organizations (NGOs) have a habit of publicizing the country's poverty and human rights violations, which the country also hates. A group of Americans behind me in customs naïvely asked what box to check if they were here to volunteer; they were quickly ushered by guards into a separate area for questioning. After an hour, my visa was finally approved. As I learned, a US passport usually means that you are bringing money into the country, regardless of your actual reason to visit; this leads to very few questions if you say as little as possible.

Upon leaving the airport, I was immediately greeted by at least 30 taxi drivers. There was only one flight arriving in Dar es Salaam that evening and they waited all night for this one business opportunity. If you are new to the country, doing such business in Tanzania can be stressful. Everything is done in cash, and all prices are negotiable. The starting price depends on what you look like. East Asians are generally charged the most, as most of them are here for business reasons and likely on an expense account. Americans and "mzungus" (colloquial term for a Caucasian) are mostly here for tourism and are charged the next most. As a reference, the ride from the airport to the city for a foreigner costs about 30-40,000 Shillings ($14-18). This amount of money is the approximate life savings of an average Tanzanian. Of course, after factoring in the cost of car maintenance, fuel, and airport fees, there is very little money left for the actual taxi driver for this ride.

Ironically, the above factors translate to high levels of social trust upon economic matters. No one ever asked for money upfront. And if I didn't have enough cash, drivers were okay with collecting money the next day. Booking tickets for rides, tours and safaris are done verbally over the phone or via text messages, with no official reservation confirmation or receipt. No one ever asked for proof of a reservation either. Yet everything works out, sometimes even more reliably than in America (e.g. think about how many copies of a confirmation and forms of ID you need for any routine reservation). Local businesses and individuals rely on word-of-mouth reputation above all and that trust lowers the transaction costs that would be typical in America. I suspect that social trust is crucial for their economy; without it, their markets would cease to function effectively.

What would you like to share about Tanzania?

Tanzania is a relatively new country. It was formed in 1964 after the merger of two separate countries- Tanganyika and Zanzibar (hence the name Tanzania). Tanganyika was once part of German East Africa before becoming a territory of the United Kingdom. Zanzibar, on the other hand, was once the former capital of Oman / Muscat, the center of the Arab slave trade, and a crucial leg of the spice trade. These historic roots have had lasting impacts. For example, over 99% of Zanzibar is Muslim, and large Indian and Middle Eastern populations have lived there

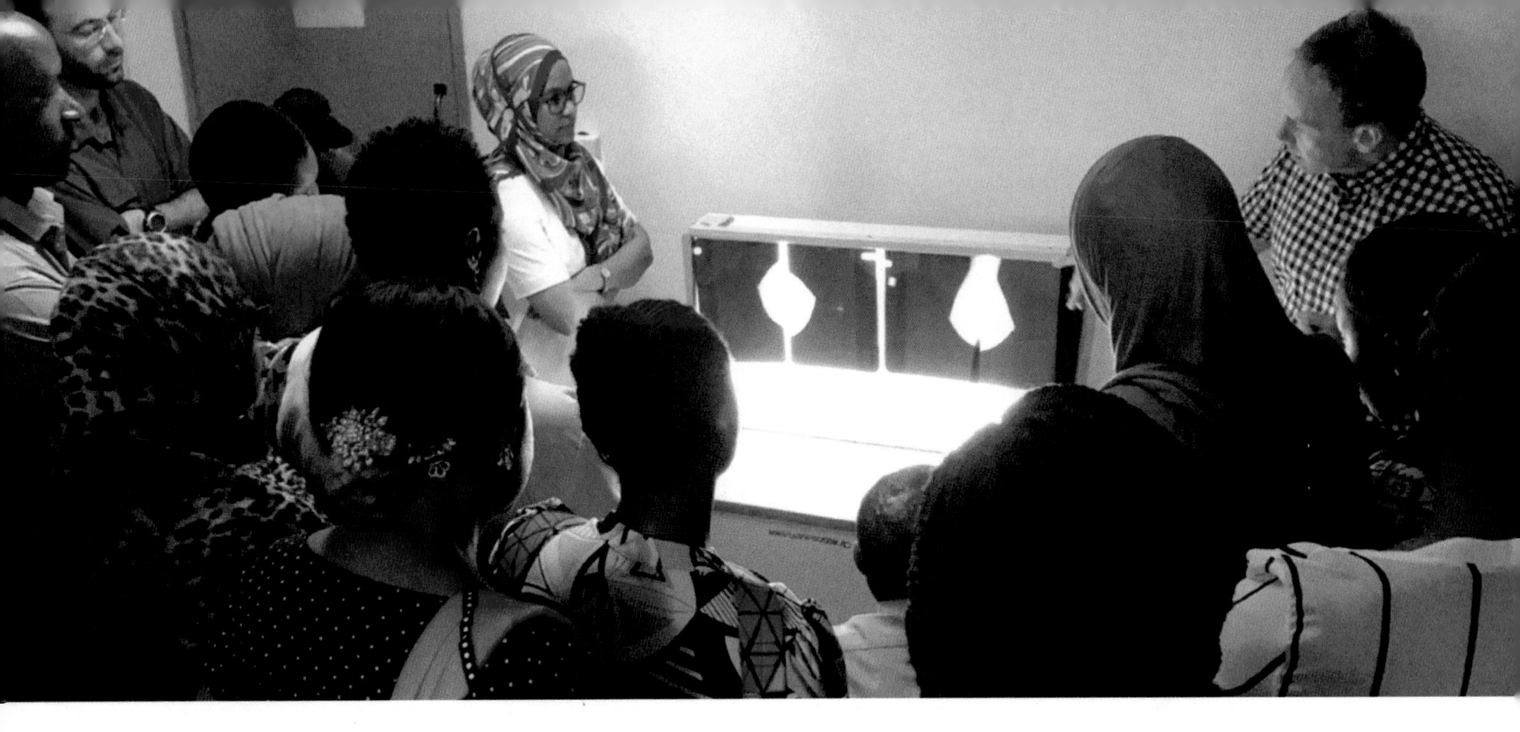

for several generations as a direct result of either the Arab slave or spice trade (in fact, Freddy Mercury was born in Zanzibar to Parsi parents from India.) On the mainland, the predominant religion is Christianity, and there are neighborhoods that are historically British (e.g. Oyster Bay) juxtaposed with local Tanzanian districts (e.g. Temeke).

The country is home to over 60 million people, but only one city in the country has over 1 million people- Dar es Salaam. The rest of the population is scattered across the country's rural interior. Although the capital moved from Dar es Salaam to Dodoma (in the central region) a few decades ago, Dar es Salaam remains the de facto center of economy and government.

If you are a local Tanzanian, the first language you learn is your local village dialect (there are 126 of these). This is what you use to communicate with your relatives and friends. Swahili, the lingua franca, is learned next. This language unites over 100 different ethnic groups in Tanzania and is spoken in neighboring East African countries, including Kenya, Uganda, and DR Congo. English is learned 3rd, and though it is the language of business (and medicine), only a minority of the population can carry on a conversation in English.

How is radiology practiced in Tanzania?

I rotated at Muhimbili National Hospital, one of four national hospitals offering tertiary level care. This 1600-bed hospital is served by just eight staff radiologists. As a comparison, I counted over 90 staff radiologists at Stanford + LPCH for 900 beds! The hospital has two MRI scanners (one inpatient, one outpatient), two CT scanners (including a 128-detector Siemens Definition Flash scanner), and four ultrasound units. Room turnaround time is quick to keep up with the high patient demand; the inpatient MRI scanner performs about 40 studies per day.

Radiology training is very different from the US. Radiologists first have to complete a 5-year medical school (considered a bachelors degree) and an intern year. Afterwards, they work as a "registrar" (equivalent to a general practitioner or senior internal medicine resident) either for a few years or indefinitely, usually in their hometown. If their hospital has a need for a radiologist, they will sponsor a registrar to complete further training in radiology residency. Because subspecialty training is considered a masters degree, residents pay tuition (typically partially funded by their registrar hospital) despite also participating in the night float and call pool. Residency training is three years; first year is mostly didactics and ultrasound scanning, while the next two years are mostly plain films, CT, and MRI. Fellowships do not exist. Following residency, the radiologists are obliged to return to their registrar hospitals.

There were several differences between how radiology is practiced in Tanzania versus the US. Imaging requisitions are never reviewed for appropriateness or for protocol optimization. Sonographers interpret >90% of ultrasound studies themselves; only a few key images are saved and printed, and never looked at by a radiologist. Plain films performed in rural clinics may never be seen by any physician. A red post-it note is used to flag a stat study. Stat, however, is relative. The turnaround time for an ED study can take up to three days. For example, on a Monday, I interpreted a CT head on a patient with an acute MCA stroke and early signs of herniation. When I asked how to reach the primary team, the other residents looked puzzled; no one had ever communicated a critical result to another clinician. This endeavor was moot anyways, as I realized the CT had been performed three days prior (on Friday afternoon).

I think part of this stems from the fact that while radiology has existed in Tanzania for decades, radiologists as a profession is still relatively new. After all, the first radiology program had only begun 10 years ago. Without a strong precedent, radiologists' role within clinical care is still undefined. Turnaround times, communications of critical findings, and protocoling studies may seem like a nuisance on call to us. On the flip side, I witnessed an environment where these duties are either undervalued or nonexistent, and it's a struggle for their radiologists establish a role in patient care.

At the workstation, I spent most of my time teaching their junior residents my search pattern and approach to different studies. This was harder than I anticipated, as their imaging approaches our complexity. Sometimes I was asked to scan a renal transplant allograft, interpret a CT coronary CT angiography, supervise a body MRI, or interpret an MRI brain spectroscopy. All of this on the fly- with no StatDx, e-anatomy, or other Internet resources available. Luckily, the Yale attendings bailed me out when I needed help.

Interestingly, it is also much more crucial to establish a diagnosis radiographically. Many of the differentials we come up with at Stanford might seem like an academic exercise particularly when a suspicious mass is getting biopsied anyways. However, it is extremely uncommon for this step-by-step workup to occur in Tanzania- core biopsy and PET/CT are unavailable at this (tertiary, national) hospital, and a patient may proceed directly to surgery or chemotherapy based on what the imaging shows.

Why should radiologists bother with global health?

According to the WHO, 2/3rds of the world does not have access to basic radiology equipment, yet at least 60% of conditions rely on imaging for a proper diagnosis (100% if you're at Stanford). To emphasize the importance of imaging in medical care, take a look at the WHO Model List of Essential Medicines, a list meant to reflect the minimum and universal needs of a basic healthcare system. Since its 1st edition in 1977, iodinated contrast made this list. You know what else made the list? Oxygen. As radiologists, I think we easily forget that we have just an equal stake in a patient's outcome as the ordering provider. When we approve a contrast study protocol, we consciously agree with clinicians that the benefits outweigh the risks, and that such benefit includes the timely diagnosis that may save the patient's life. If the WHO released a list of essential equipment, I'm willing to bet that an x-ray or CT scanner would make the top of the list. And if they released a list of essential specialties, radiologists would be in the top two–tied for whichever specialty is tasked with ordering the study.

At the same time, the presence of radiology equipment doesn't equal radiology. I think there is a misconception that global health radiology can be accomplished by installing modern CT/MRI equipment in a developing country and then sending the images through the cloud to be interpreted by radiologists in developed countries. That misses the point of what a radiologist actually does. We are responsible for designing protocols, ensuring that these protocols produce quality images, assigning the correct protocol for the indication, communicating critical findings to providers, and helping clinicians determine the next step in management. At a broader level, we are also advocates of public health (e.g. breast cancer, HCC screening), stewards of just resource utilization (imaging wisely), and experts of radiation/contrast safety. I would thus argue that image interpretation by itself constitutes a very small fraction of what we ought to be doing to maximize impact.

Justin Tse, MD

PGY-4 Resident
Stanford Medicine | Radiology

We Must Reject Stereotypes to Promote Diversity

One of the most powerful features needed to promote diversity is to persistently reject the notion of what some would like to call "stereotypes" or what Chimamanda Ngozi Adichie, a prominent award- winning Nigerian writer, once called the "Danger of a Single Story" [1]. From her own life's experiences, she weaves true anecdotal evidence that highlights the dangers of looking at ourselves, others or a group of others through a single lens. As a tribute to black history, we would like to highlight a few black characters or heroes who defied tall odds and shattered the mold of stereotypes, to make several critical contributions to STEM, medicine and education, rejecting the status quo in the process through hard work and perseverance to make history that opened new doors to their gender, cultural and racial background.

Most people in the U.S. and around the world probably know of Martin Luther King Jr. [2], renowned for his non-violent but powerful activism that broke many stereotypes. He led the civil rights movement in America which was monumental in securing equal rights for blacks and other minority groups. He also famously said: *"Cowardice asks the question, is it safe? Expediency asks the question, is it politic? Vanity asks the question, is it popular? But, conscience asks the question, is it right? And there comes a time when we must take a position that is neither safe, nor politic, nor popular, but one must take it because it is right"* and that *"An individual has not started living until he can rise above the narrow confines of his individualistic concerns to the broader concerns of all humanity."* Therein lies the heart of what it takes to reject stereotypes in today's world and to be open to diversity and inclusion.

In Engineering, we read heroic stories like that of Dr. Shirley Ann Jackson [2], a pioneering African American woman whose career is marked by many historic firsts including first African-American woman to receive a doctorate from Massachusetts Institute of Technology (M.I.T.) in any subject; one of the first two African-American women to receive a doctorate in physics in the U.S.; first African-American to become a Commissioner of the U.S. Nuclear Regulatory Commission; first woman and the first African-American to serve as the chairman of the U.S. Nuclear Regulatory Commission; first African-American woman to lead a national research university i.e. eighteenth president of Rensselaer Polytechnic Institute; first African-American woman elected to the National Academy of Engineering. Her story should inspire young URM ladies with tall dreams, to know that they indeed can walk where no one like them has walked before.

Another groundbreaking, innovative engineer and a fellow of the US National Academy of Inventors, Dr. Thomas O Mensah [2] from Ghana inspiringly states in his autobiographical memoire: "The Right Stuff Comes in Black, Too." Dr. Mensah is a world-renowned inventor of fiber optics technology with several pioneering patents awarded within six years. His work is essential to the spread of fiber optics networks throughout the U.S.A. and for the global reach of internet technology platforms through ultra-high strength optical fibers for terrestrial and submarine cable transmission applications. His work is critical for the modern-day fiber-optic communication necessary for fast transmission of pictures, videos, and other forms of data worldwide. Dr. Mensah is also passionate and intimately involved in ensuring the earth's sustainability and in advancing and unlocking the numberless promises that nanotechnology holds for mankind.

Mae Carol Jemison has a diverse plate of careers including astronomy, chemical engineering, education, and dancing [2]. She was the first African-American woman to travel to space. She excitedly danced her way into space aboard the Space Shuttle Endeavour on September 12, 1992; spearheading a novel exploratory path of black excellence for many who look like her, as they were thrust into space. She is a showcase of what is limitlessly possible when we stubbornly dream in active pursuit of the same. She brings her passion, enabling the interaction of science and technology with our humanity and elevating prospects of having minorities walk through doors, which only a few years ago would be closed to them by racial prejudices and racial biases.

Another inspiring story, in the field of Radiology is that of William E. Allen, Jr., MD, FACR [3], a pioneering physician and leader, who arrived for his American Board of Radiology (ABR) examination in 1935, and was directed to take the freight elevator because the hotel manager said African-Americans could not share one with the white guests. Allen rode to the top floor, took the oral exam, passed it, and became the first black diplomate of the ABR. He later became the first black member of the American society of Radiology (ACR), the first black fellow in 1945, and an ACR Gold Medalist in 1979. He has also developed scholarships for students from Haiti, Nigeria, Liberia, and South Africa to study radiology.

[1] https://www.ted.com/talks/chimamanda_adichie_the_danger_of_a_single_story#t-778995
[2] http://www.blackheroism.com
[3] Oestreich AE. William E. Allen, Jr, MD, FACR: pioneering physician and leader. J Natl Med Assoc. 1999 Jul; 91(7): 414–418.

Growing up beyond fear and doubt

Mario Moreno's family came to the US as illegal immigrants. Poor grades in college shattered his hopes of getting into med school. After undergrad, he worked as a night time interpreter in a hospital and studied for his MCAT between patients. He was feeling doubtful and afraid. Listen to this extraordinary story of determination and recovery to achieve the height of success in medicine!!

https://soundcloud.com/user-333680609/growing-beyond-doubt-and-fear

Mario Moreno, MD

Radiology Resident
Stanford Medicine | Radiology

Insights from a South-American-Latino-Jewish European-US citizen

I grew up in Argentina, where hair color and skin tone inherently stratify humans into different ranks that then define beauty standards, as well as who gets what job, who is feared and who is automatically accepted, who gets a table first at a restaurant or who is allowed into a nightclub.

I am Argentinian. I am South American and Latino. I am Jewish. My grandparents emigrated from eastern Europe before the second world war, escaping the Nazis. My skin is not quite "European" white, nor its black. I am of uncertain origin - a light brown color - that has never fully allowed me to fit in one specific skin color box. Darker people see me as white and whiter people see me as dark.

As a South American-Latino-Jewish European-US citizen male with light brown skin and a thick accent, I have often felt the victim of unconscious bias. I have often wondered if I'm deemed inferior because one or many of these qualities? These feelings are often hard to justify because the actions that caused them are essentially subtle, unconscious and most of the time unintentional.

In this country, my story is one of privilege. While I am empathic with the cause of bringing freedom, safety and prosperity to the black community for once and for all, it is impossible for me to fully understand the racial suffering that many African Americans have sustained since the beginning of this nation and even before that. This suffering continues today and is evident by looking at every health statistic of the black population of this country as well as the many acts of explicit violence and discrimination that we all witness frequently.

I know that the United States is a developed, high-achieving nation with the primordial intention of generating equal opportunities and erasing structural differences and racism. This ideal has at least in theory made this country stand and rise above others. But there is still a long way to go to make this a reality and, frankly, the time for waiting is done. We must work harder every single day to achieve this, to put these deep inequities in our past and in the past of the country and, most importantly, of our children.

Guido A. Davidzon, MD, SM

Clinical Associate Professor
Nuclear Medicine and Molecular Imaging
Stanford Medicine | Radiology

Echalé ganas, Mijo

"Echalé ganas, mijo," My grandpa would often say — apply yourself and put efforts into your studies. Throughout my childhood in California, this phrase was ubiquitous; I'd hear it at Mass, at the taquería, and certainly at home. Whenever I met family members for the first time, I came to expect these three words.

I'm one of the lucky few. Whereas I received financial aid to attend both a private prep school and now an Ivy League university, many students in East San Jose are simply told to "stay in school." There's a world of difference between that sentiment and "echalé ganas," though on the surface they seem similar. Students from all backgrounds, but especially those who have too often been left behind in a system that has historically favored the white and wealthy, should not merely be staying in school — they must be thriving in school. Read the full story here: http://hechingerreport.org/student-voice-we-must-offer-underprivileged-students-more-than-a-meager-stay-in-school/

Cayo Gonzalez

Former high school student in Dr. Sanjiv Sam Gambhir's lab (2016-2017).
Currently, an undergraduate student at Columbia University.

Not everyone can be in the front-line taking care of patients or developing new vaccines during the COVID-19 pandemic. I wondered how my engineering background could complement ongoing efforts of our clinical colleagues. In this direction, my solu-tion focused on improving airborne sampling methods towards noninvasive virus collection and reducing false negative test results. I was fortunate to receive support from the "Stanford RISE COVID-19 Crisis Response Train-ee Seed Grant Program" which helped to fast track the development of my work. Thus, in just a few months, my colleagues and I translated a new idea into a practical solution. As part of Stanford's School of Medicine, we are the privileged few that can significantly impact our community and the world. Our innovation spirit will lead to many engineering and medical advances that will be passed on to future generations. In the dawn of a new year, I look forward to new beginnings. There is work to do, and that's why we are here.

Fernando Soto Alvarez, PhD

Postdoctoral Fellow
Canary Center for Cancer Early Detection
Stanford Medicine | Radiology

"You are not lucky to be here.
The world needs your perspective.
They are lucky to have you.

— Antonio Tijerino

President & CEO of the Hispanic Heritage Foundation

"Don't wait until you die to be reborn."

Mayra Rubio

"Change does not roll in on the wheels of inevitability, but comes through continuous struggle."

Dr. Martin Luther King, Jr.

"Even if you are a minority of one - the truth is the truth."

Mahatma Gandhi

"Never be limited by other people's limited imaginations."

Dr. Mae Jemison, first African-American female astronaut

WOMEN IN STEM

"Let us choose for ourselves our path in life, and let us try to strew that path with flowers.

– Emilie du Chatelet
mathematician, physicist, author

Women in STEM

This chapter is dedicated to women in Science, Technology, Engineering and Mathematics / Medicine (STEM). It sends a note of gratitude to the many female physicians, researchers, staff, and trainees in the STEM field for their dedication, perseverance, inspiring ideas and hard work. This chapter will provide examples of unique talents and accomplishments of female trainees and faculty in the STEM field, with a specific focus on medical imaging. We will also learn about obstacles and roadblocks that women in STEM still face today. And we will learn how a little support can make a big difference.

For many women in academic medicine, work experiences are like a roller coaster ride – more extremes than in most other professions. Every one of us has encountered obstacles, setbacks and hurtful disappointments, sometimes from a direction we least expected. No matter how we feel, we get up, dress up, show up and do not give up. We are grateful for our supporters, both male and female, who are helping us along the way. It is an honor and pleasure to celebrate our successes with you!

We are optimistic that the underrepresentation of female faculty in the fields of medical imaging, engineering and computer sciences will improve in the near future, and that the current gender imbalance at higher ranks will vanish in one generation. But this will not just happen. We need your help to initiate change. Everyone in the STEM field needs to help and actively support both male and female trainees and colleagues in order to enable them to realize their full potential. Our community will benefit from their groundbreaking ideas and discoveries!

Thank you for supporting women in STEM!

How to be an ally

Women in STEM have not made it through the glass ceiling yet. While many academic institutions now train close to 50% female students in STEM disciplines, there is a major underrepresentation of qualified women in leadership roles. As a community, we must come together to spot, support and advance women in science and medicine. We need to seize the full breadth of our nation's scientific and economic potential. The suggestions below provide examples, how women in STEM can be supported. These points are not meant to be comprehensive and may require adjustments for different communities and contexts:

- Encourage young girls in your life, such as your daughters, cousins, nieces and/or your friend's child to explore an interest in science and STEM careers
- Create and support outreach and pipeline programs
- Enable women in your community to share their insights and ideas. Listen.
- Create a code of conduct to clarify appropriate and inappropriate behavior in your department / institution.
- Amplify the voice of your female trainees and colleagues: If she contributes something, affirm her statement: "This was a great idea..", "I agree with Susan.."
- Measure male and female speaking times at meetings
- Introduce objective, unequivocal, quantifiable metrics for annual evaluations and career advancements. Measure the reproducibility of these metrics.
- Define rules and criteria for candidate selection (appointment), candidate assessment (career advancement) and creative work evaluation (manuscripts, grants) before any evaluation takes place
- Add a diversity advocate to appointment and promotion committees
- Promote, support and publicly recognize research conducted by women
- Speak up and intervene if you see verbal or physical harassment
- Evaluate and address gender disparities of compensation and promotion data
- Clearly communicate formal procedures for reporting verbal or physical harassment to all Department members
- Introduce formal procedures to report microaggressions against women, such as sexist language or jokes, undermining assertive women by labeling them with sexist stereotypes ("bitchy", "bossy", or "aggressive") or devaluing women by talking over them and co-opting their ideas
- Link leadership training to assignment of a leadership role. In analogy, if someone attended driving school, they would be expected to drive a car afterwards. Too many women were sent to leadership training courses and never realized their potential as leaders.
- End lifetime leadership assignments and fiefdoms in medicine. Introduce rotating leadership roles. Introduce a voting system for leadership roles.
- Introduce prospective annual performance meetings: "next month/year, you are expected to do x", be specific about x, then review if x was achieved.
- Leaders: Role-model professional behavior that avoids sexist remarks and culture
- Monitor female-to-female behavior. Incentivize female support personnel supporting female leaders and vice versa.
- Link incentives of leaders to the success of their team members
- Academic centers: Increase endowed professorships for women
- Assign time for mentorship, incentivise and reward participation in mentorship programs for both mentees and mentors
- Organize save spaces for frank discussions, such as a monthly womens networking lunch or dinner

How I Define Success

Reflections on how to succeed as a woman in radiology, that is my task. "How to succeed" is the difficult part as the word "success" can have so many different meanings. My definition of a successful career is one where I can do meaningful work that I enjoy, make a difference for my patients, and still be present for my family and friends. Success in your career is not a contest that is either won or lost as there are many wins and losses along the way.

I was very fortunate as a medical student and resident to have the late Dr. Helen Redman, past RSNA president, and SIR Gold Medal recipient, as my mentor and sponsor. Even long before those words were really used, she took an interest in me and helped push me into the field of interventional radiology. As the first medical student allowed to rotate in IR, she really opened my eyes to the possibility of a career that was almost surgery - but better (my husband was a surgery resident, so I knew that for sure). In a national interview, she called out women who decide against a career in IR after having children as a serious loss to the field of radiology. Needless to say, I forged on in IR, as I was not going to let her down, having three children along the way.

Most women still do not choose my field, with less than 10% of practicing IR physicians being female. Reasons for that are unclear. However, women are reported to gravitate towards "patient-facing" careers. We are definitely patient-facing and are known for doing innovative "cool" cases, however, in pediatric IR the "cool" case may be on a one- -week-old or a 13-year-old, which changes the entire dynamic. We are both patient and parent facing, and the ability to be empathetic and compassionate for our patients and family, yet still innovative in treatments, is one factor that drove me to switch from the adult to pediatric world. This mid-career change was influenced both by the need to feel in control of my time as well as the search for the feeling of being appreciated. Yes, I definitely met the definition of physician burnout. Radiology is such an amazing field that one can actually completely redirect their career along the way, as long as the drive for discovery and learning is present.

My keys to success: be passionate about what you do, seek meaning in what you do, stay empathetic. There is a patient behind that X-ray. Do not forget, no one will remember you for how fast you can get through the stacks or how fast you can do a TIPS, but they will remember you for how you treated the people you work with and your patients. Your family will only remember how you treated them, and that is where I define success.

Shellie Josephs, MD

Clinical Professor of Radiology
Director of Pediatric Interventional Radiology
Stanford Medicine | Radiology

Embracing Scientific Diversity

As I'm writing these sentences, we are celebrating the "International Day of Radiology," and I can't help pondering at how my status as a minority research scientist has influenced my contribution to the field of radiology. My "humble" background – raised in eastern Europe in a blue-collar family, as a first-generation high-school graduate – has taught me many valuable lessons, including that hard work and perseverance, are necessary but not sufficient to succeed.

A fearless attitude and willingness to step outside my comfort zone are the most important factors that shaped my contribution to science. They influenced many of my personal career decisions: choosing computer science high-school training without ever seeing a computer in person; getting an engineering degree in a country that was 1500 miles away from my home town; changing fields between my Ph.D. and postdoctoral training; transitioning from academia to industry and back to academia. Scientifically, these experiences reinforced my simultaneous interest in multiple fields, placing my research at the boundary of multiple biomedical and technical disciplines with very diverse applications. For example, my Ph.D. research has been focused on characterizing protein structures at multiple scales using computational approaches, while I currently use this knowledge to spatially correlate pathology images with radiology exams with the goal to better characterize diseases and cancer aggressiveness. By bridging the gap between multiple modalities, my research addresses their individual limitations and benefits from their joint advantages.

When working with the younger generation of radiology researchers, I feel it is my responsibility to emphasize the need for a diverse research training that spans many fields, for example, computer science, medicine and mathematic- or biomedical-research, precision health, and precision medicine. Diversity has many faces. Embracing scientific diversity helps us be better researchers.

Mirabela Rusu, PhD

Assistant Professor of Radiology
Integrative Biomedical Imaging Informatics at Stanford
Stanford Medicine | Radiology

Be Like Water

Female radiologists are still a minority in a male-dominated discipline. We are underrepresented amongst radiology residents, fellows, junior faculty and radiology leadership. Many of our parents wanted our brothers to become physicians and us to marry one. While many people we interact with are supportive of female physicians, patients, students and colleagues frequently confuse us with nurses, technicians or administrative assistants. When we correct this, we get mixed responses. We have to explain that we are not disregarding anyones importance. We are correcting a stereotype. Our gender does not define our role in the health care system.

It is important to understand that many women who enter the field of medical imaging have to deal with daily criticism and microaggressions: Our colleagues provide (perhaps well meant) unsolicited "advice", our trainees scrutinize our appearance, our patients question our competence and all of the above score us lower on evaluations than our male peers, typically without rational explanation. I recently learned that this phenomenon has been well described in the social sciences: It is called the "tall poppy syndrome" (google it). It refers to a culture in which people of genuine merit and conspicuous success are resented, attacked and undermined with the goal to bring the "tall poppy" down. The tall poppy can be a successful woman or another high achieving individual who is perceived as "different".

A typical reaction by junior "tall poppies" is to try to blend in, to avoid speaking up and to avoid standing out. It took me many years to realize that this might not be the best approach and that this entire reaction is not about me. It is about people who are unhappy about their own career choices or who feel threatened somehow. They will criticize you no matter what you do. This is actually very easy to address: Surround yourself with supportive people – people who encourage you to be the best you can be and who are happy to see you grow. Be like water: If barriers enter your life, just flow around them. Reprogram your brain with the confidence that nobody can stop you. Be comfortable with being unique and strong. Our daughters are watching you. And we will all benefit from your success!

I am grateful that I found my supportive community at Stanford Radiology!

On my flight on my way to take the ABR certifying exam, as I was reviewing the non-interpretative skills packet with my baby sleeping soundly on my chest, I heard the dreaded overhead announcement: "Is there a doctor on board?" As a radiologist, I always worry it's going to be a real medical emergency requiring me to run a code or something. I put the call light on and the flight attendant came over and immediately started speaking to my husband, assuming he was the doctor. In any case, I had the honor of assisting an elderly lady with shortness of breath. I took her vitals, listened to her heart and lungs, and gave her some oxygen, water and orange juice (for low blood sugar). Thankfully, she seemed stable enough to make it to the landing, and didn't have more serious or progressive symptoms! That was a little unexpected excitement in my day. So much for studying on the plane - but it feels good to be able to help in a situation like that!

Crystal Norgren-Farrel, MD

Pediatric Radiology Fellow
Stanford Medicine | Radiology

Same thing happened to me! I pressed the button and the flight attendant said: " I can't help you now. We have an emergency! "

Karen Ordovas, MD

Professor of Radiology
University of California San Francisco

The Impact We Make Depends on Our Principles

As a PhD candidate in engineering, I had five tenured professors on my dissertation committee. They all positively contributed to my development into the strong scientist I am today. They taught me the value of logic, rigor, integrity, innovation and perseverance. They taught me to work hard and to play hard, to love my work and to live my life. They taught me to fight for the principles I believe in and to believe that change is possible, no matter what.

All five of my dissertation committee members were male. There were no female mentors available to me in my academic bubble at the time, and consequently it never occurred to me to proactively search for any during those deeply formative years. The lack of options back then indicates that gender diversity and awareness were indeed missing – but even as a minority female engineering student, I didn't see how this might affect me in the future. Nevertheless, my committee instilled within me the confidence to become a leader and independent thinker, and I sailed off into the post-graduate horizon with only the belief that I would succeed.

As I establish my own career path and see the disheartening statistics facing women and minorities moving up the academic ladder, I realize that diverse mentorship is needed today more than ever. I now think about the pivotal concepts I want my students to remember. Naturally, I want to pass on the fundamentals of good science that I've learned from all of my teachers thus far. Additionally, I want to share equally important lessons that I've learned from postdocs, staff and faculty at Stanford – women and men who have come from different parts of the world, who have found ways to triumph against the odds, and who have unknowingly become non-traditional mentors in sharing with me their personal stories about trial and error, success and failure. To move toward a more inclusive and collaborative system of science, we must seek out and serve as mentors who implement human fundamentals of empathy and equality into our everyday interactions. Otherwise the scaffolds of our cherished academic world would surely fall apart.

The impact we make on future generations depends on the principles we choose to cultivate and our availability as mentors to our peers and colleagues of any age, race, or gender. We should all remain students and teachers for as long as we can, both in research and cultural settings. We should openly propagate awareness about our academic values and challenges, and actively listen to the voices of our neighbors. Importantly, these actions will ensure that strong scientific leadership is sustainable, for women and men alike, and this will encourage all scholars to become mentors who can make a meaningful difference too.

Sharon Hori, PhD

Instructor of Radiology
Molecular Imaging Program at Stanford
Canary Center at Stanford for Cancer Early Detection
Stanford Medicine | Radiology

Challenges for Women in Academia Are Real . . .
and Increasing Awareness Will Help Us Grow

I prefer not to think of myself as a "woman in radiology," but rather a basic scientist in radiology who happens to be a woman. However, regardless of what you call it, there are some unique challenges that women face in academia. Despite working in an area of science that is heavily male-dominated throughout most of my training, I remained largely oblivious to many of these challenges, but over time they have become apparent and should be recognized and discussed rather than minimized and dismissed. Here I would like to describe some of my experience with various challenges that women face and the associated growth opportunities.

One of the first things that I realized when starting my lab is that direct communication from women is often perceived unfavorably. This is something that I had never thought about but realized quickly when I started my lab. When I made the transition from an individual contributor as a postdoc to a group leader as a faculty member, it became evident that women are often perceived as bossy or mean when they speak candidly. While I have learned to cultivate my communication style over time, the real growth has come from reflecting on my own reactions to how others communicate. A second challenge that I have faced is the difficulty that many women share in promoting themselves and their work. Although some aspect of this is personality-driven, I have noticed in conversations and scientific talks that women tend to use "we" much more frequently than "I" when describing their work. In lab and other meetings, I have made a deliberate effort to encourage speakers not to do this. While this kind of modesty may prove to be a difficult thing to change, one thing that is very easy for us to do is to promote the work and accomplishments of each other, as a way of drawing attention to achievements. A third obstacle that I have seen women face is in how they are described by others. Instead of focusing on scientific skills and accomplishments, many recommendation letters for women are often filled with descriptions of how well they got along with their colleagues or descriptions of other interpersonal relations. Although these points are important, I have made a deliberate effort to focus on the scientific aptitude and qualifications of a person when writing and reading recommendation letters. And finally, a force that other women and I contend with is the biological clock. The inherently limited window in which women can typically have children often coincides with critical points in a scientific career at the end of training and the beginning of an independent career. While this is a very personal decision, many people realize that if we wait until the right time to have a child, we may never find the right time. My husband and I welcomed our first child this past summer, and I have realized the pros and cons of having a baby at age 39. While having Baby Nolan has been life changing in such a positive way, the physical and mental effects of motherhood are real. I have found talking with other women (and men) to be immensely helpful in finding the needed support. Now I see a tremendous opportunity to pass this support on to other women.

Of course, not everything that I have described here is limited to women, and many of these challenges are multifactorial, but they highlight some obstacles that should not be ignored. And while every situation is obviously not immediately solvable, I believe that increasing awareness will allow us to grow. We are so fortunate in this department and at Stanford to have tremendous female role models who are very open to sharing their experiences. One of the most powerful things that I have done here and throughout my career is to find strong mentors (both female and male) to help navigate various issues. I am very grateful for their perspectives and am happy to pass this gift along.

Sharon Pitteri, PhD

Associate Professor of Radiology
Canary Center for Cancer Early Detection

Insights from Summer Student Irmina Benson

Irmina Benson, a summer student at the Canary Center, shared her experiences on CBS/KPIX:

"My siblings and I did face a lot of discrimination and a lot of racialized bullying when we were younger," recalled Benson. "That was very difficult."

Studying became Benson's safe space. Benson plans on being a doctor, and serving communities of color and others who need extra care. Her training at the Canary Center at Stanford is part of that future.

https://sanfrancisco.cbslocal.com/2018/11/06/following-mothers-cancer-diagnosis-bay-area-student-pursues-career-in-medicine/

Lab Life – When Privileged Are a Minority, Equity Stands a Chance

Amin Aalipour, MD, PhD student and member of the Gambhir lab, shared his experiences in a recent opinion article in Nature:

"In my experience, equity prevails in groups with no apparent ethnic or gender bias. I have been a student in six labs and collaborated with countless others. My current lab is the most heterogeneous: of 37 members, 14 of us are female and our backgrounds span 17 countries. Only two fit the description of 'white American male'. I find it empowering to work in a gender-diverse, multicultural environment that is quick to rebuke entitled behaviour."

Nature 563, 473 (2018)
doi: https://doi.org/10.1038/d41586-018-07453-3

Stanford Radiology summer student Kensley Villavasso is not only a gifted scientist but also an accomplished artist. Kensley has been drawing for 15 years, and her work was showcased in galleries, at plays in New Orleans, and the New Orleans Museum of Arts. Kensley generously shared some of her pieces with us.

Top - ***Double Vision***
I was motivated by the thought of every person having different sides to them that not everyone can see.

Bottom Left - ***Clarity***
This a multimedia piece on canvas. This piece is a part of a collection I did in 2016 depicting different parts of the face and mind.

Bottom Right - ***Expressions***
This is a color pencil piece about internal feelings. I was motivated by feelings that we are told to suppress and bottle in until we cannot hold them back anymore.

The Martial Arts of Respect, Discipline and Humility

Kenpo Karate is my hidden talent. I began studying Kenpo shortly after finishing my radiology and nuclear medicine train-ing. As I started my journey as a physician scientist and nuclear medicine attending I wanted something that could help me feel more focused, disciplined, confident and empowered.

Kenpo Karate takes its roots in the Chinese martial arts, with heavy influences from the Japanese martial arts systems. This traditional martial arts system maintains many of the classical elements of training, while incorpora-ting practical and tactical elements for use in a more modern self defense environment.

I joined the local dojo by my home, and began practicing with my Sensei 4 to 5 days a week. Training always begins and ends with brief meditation during which time we focus on what we have learned, what we have done well and what we wish to do better. I spend my training sessions either working on forms, defense techniques with a partner, sparring or weapons training. Currently I am studying nunchuck techniques with the goal to ultimately master the shaolin whip and chain whip. The school that I train with is the Z-Ultimate Self Defense Studios. We practice with the three core values of Re-spect, Discipline and Humility. These values are defined in our course work as follows (in abbreviated form):

Respect: Many people confuse respecting someone for fearing someone or automatically treating that person as better than you, as if they are deserving of more. This is completely wrong. Respecting someone – including your-self – means valuing that person; recognizing their worth as an individual and appreciating not only what they can do on and off the mat, but what they are capable of doing.

Discipline: In life, many people say that they want to achieve great things. What separates those who are masters of their lives from those who are dreamers is that the masters make themselves do what they said they wanted to do, regardless of current circumstances. Mastering the martial arts – as with anything in life – requires constant repetition, or the "habit" of practice. Habits are only developed through focused behavior towards a goal.

Humility: Humility is not about thinking less of yourself; rather, it is about being thankful for how unique and valu-able you are as a person, and understanding that you have a responsibility to share your gifts with others. The greater the martial artist, the greater the responsibility to pass on the knowledge to others. A true master realizes that it is not about the individual; it is about what the individual stands for."

The Martial Arts of Respect, Discipline and Humility

I have found through my training that I am more focused, more confident and better equipped to pursue my goals inside and outside the dojo. As a woman in Radiology and Nuclear Medicine, I have found my practice of martial arts to help me feel more empowered to be a pillar of strength for other women and minorities facing professional challenges. I recently prepared for and participated in my first tournament – the 2018 Winter Grand Championship. To me, choosing to compete was a true test of conquering my insecurities and finding my own sense of power. I prepared diligently, and was thrilled to take home first place trophies in both Sparring and Weapons for my division.

I am currently at the rank of a purple belt, and have enjoyed my journey in the martial arts so far. I look forward to continuing to train, learn, practice and compete. I feel strongly that the skills I have learned in the dojo have helped me become a better physician scientist for my patients, a better leader for organized medicine and a better mother for my three boys.

K. Elisabeth Hawk, MS, MD, PhD

Clinical Assistant Professor
Stanford Medicine | Radiology

The Art of Verbal Communication

As a liaison in Pediatric Radiology, there is a vast array of offices, nursing staff, surgical staff, schedulers, front desk staff, and numerous departments and physicians, all needing information regarding the patient at the center of their world at any given moment. While patient information is available in a variety of formats, from charts, graphs, images, historical foundations and test results, etc. the art of verbal communication and expression can convey the attitude of personal care that touches the heart and mind to give confidence and concern. With the rise of artificial intelligence and mechanical knowledge, the faith in medical experience and constructive listening can become missing components in our values and mission.

Behind the scenes I enjoy gifting each caller with my personal touch of a warm welcome on the phone and a smile in person, actively listening to their needs and helping them to find the needed radiologists and staff, hopefully to provide a better experience and better care for their patient. Each patient could be a neighbor, family member, friend, or co-worker who has a need for medical care. I decided almost 35 years ago to bring this essence of patient care to not only my every day life but also in dedication of my work here at Lucille Packard Stanford Children's Hospital. As a mother of two wonderful children and 2 glorious grandchildren it was and continues to be one of my values I bring to this department and into my world of love and caring. I am Annette Scott, Pediatric Radiology Reading Room Assistant, mother, grandmother and I enjoy my work.

Annette Scott

Lucile Packard Children's Hospital
Reading Room Assistant
Stanford Medicine | Radiology

Cultural Differences in Scientific Research

My name is Wei Wu. I completed my undergraduate education in China and I completed my PhD degree in Neuroscience at the Technical University of Munich, Germany. Currently, I work in the lab of Prof. Heike Daldrup-Link as a postdoctoral fellow.

During my ten-year study and journey through three continents, I have gained not only academic experience but also lots of friends from different countries. I feel that cultural diversity sometimes can subtly influence the way people think and act. The living pace in China and the U.S. is relatively faster than Europe. I heard the phrase `Publish or Perish` from one of my friends the first day I came to the U.S. There is no doubt that the result-oriented way stimulates people to work more efficiently and effectively in North America. Comparably, my colleagues in Europe pay lots of attention to details of experiments and the arrangement of the lab. They tend to believe that a gentle pace of research with lots of repeated experiments ensures the quality and reproducibility of the experiments, which does make sense. Another difference is that the recreation and work are integrated in China but more separated in Europe and the U.S. When I studied in China, we usually took a longer break in the dormitories at noon or enjoyed a good supper with friends during the waiting time of the experiments in the evening. The reason could be that the dormitories in China are normally located inside the campus and easy to reach. It took me only a few minutes to walk from my dormitory to the lab. But I prefer a more compact working style during my overseas study because the commuting always takes more than half an hour by bike or by train.

The culture in different countries exert numerous effects on me. Prof. Xinrong Wu showed excellent multi-task processing ability, which influenced me to work efficiently. Prof. Jürgen Schlegel brought me the keen interest in scientific research, he let me know WHY is the most fascinating part of science. Prof. Heike Daldrup-Link enables me to get a clear motivation to do research since translational study is closely related to patients´ wellness. From my observation, there is one thing in common of the scientists who make great contributions. They normally have strong motivation and compassion, and enjoy what they are doing. I come to realize that publication should be a byproduct of strong curiosity and interest instead of an objective of research. I sincerely wish that I could be one of them in the future.

Wei Wu, PhD

Postdoctoral Fellow
Stanford Medicine | Radiology

My Wife in Radiology

Being married to a woman in radiology is great. Gloria practices body IR at Stanford and I practice Neuro IR at UCSF. As I quip: "she does neck down and I do neck up; between us we've got it all covered." We met as freshmen at Harvard in the most romantic setting: organic chemistry class. Once the sweet smell of toluene had cleared, we both decided to become doctors and, eventually, interventionalists. In the convergent evolution of our careers, we've come to understand each other's responsibilities to patients, colleagues, and trainees. That helps us stay in good spirits despite days that start before dawn and nights that sometimes require operating on the sickest patients in the hospital. We've developed the necessary icy determination that performing emergency procedures on guts and brains require. it's not just because of her cryoablation work that Gloria's IR clinical fellows honored her with the moniker of "Ice Queen," complete with photoshopped Disney princess Elsa's white-blond locks added.

Faced with life-threatening crises as our professional norm, approaching other problems with equanimity is a relative breeze. I'm pleased that Gloria has recently turned her considerable skills in performance improvement—focused on her husband and daughters for years—to the clinical operations of Stanford Radiology. Not a day goes by that Gloria isn't reading a leadership manual, listening to a management podcast, or attending quality meetings. Having been at Stanford since starting medical school in 1996, Gloria's learned the ins and outs of the institution and seems exceptionally perceptive of the needs of everyone from technologists to students to housestaff to faculty. Her myriad conference calls this year regarding the PACS rollout have made me grateful for my own ignorance of such matters. She was the first person I'd ever heard of identified as an Epic "superuser"—her facility with the electronic medical record seemingly has made her the go-to professor for just about everyone. She's our IT guru at home, too, much to her chagrin.

Gloria's a California girl, born in the OC and grown up in Silicon Valley. (Do remind her of her OC roots when you see her: she'll love you for it.) With our hectic schedules and the demands of on-call pagers, we're grateful to have our parents nearby. Harmonizing our work and meeting commitments has always been a challenge, and not having to worry that our children will have no one at home with them when the nanny leaves has been invaluable for everyone's sanity. When our girls were in preschool, having grandparents within a few minutes was invaluable, relieving a bit of our professional guilt at not spending unlimited time with our children. The equation is now shifting, and our parents' health issues are coming to the forefront as our girls are now 9 and 13.

Gloria and I are both younger children, a point our put-upon elder daughter reminds us of at every opportunity. Our counter narrative—which we learned in a Harvard undergraduate moral reasoning course called "justice"—is that first children benefit from a disproportionate amount of parental attention, hence their over-representation in elite educational institutions as compared to their forgotten free-range younger siblings. We younger siblings have to stick together; after all, 700 out of 900 students in Justice were first or only children.

Gloria and I are grateful to have returned to California to serve the communities we grew up in and to enable our girls to grow up here too. The irony of having gone 3000 miles away to meet someone who grew up 30 miles from you is not lost on us. But chance favors the prepared mind! We share a passion to improve our institutions, the exemplars of public and private medical education, research, and care in the Golden State. And we're proud of working in departments that can attract the best and brightest from around the world to make medicine in the Bay Area even more outstanding. Californians are the future… and always will be!

Steven Hetts, MD

Professor of Radiology
University of California San Francisco

"A woman is like a tea bag - you can't tell how strong she is until you put her in hot water."

– Eleanor Roosevelt

"We cannot expect that in the foreseeable future women will achieve status in academic medicine in proportion to their numbers. But if we are to start working towards that goal, we must believe in ourselves or no one else will believe in us; we must match our aspirations with the guts and determination to succeed; and for those of us who had the good fortune to move upward, we must feel a personal responsibility to serve as role models and advisors to ease the path for those who come afterwards."

– Rosalyn Sussman Yalow

medical physicist and co-winner of the 1977 Nobel Prize in Physiology or Medicine

"Life is not easy for any of us. But what of that? We must have perseverance and above all confidence in ourselve . We must believe that we are gifted for something, and that this thing, at whatever cost, must be attained."

\- Marie Curie

physicist, chemist, and winner of the 1903 Nobel Prize in Physics and the 1911 Nobel Prize in Chemistry

Notes

"Some people make you laugh a little louder, your smile a little brighter and your life a little better."

– Anonymous

LGBTQ+

Our next chapter is dedicated to our Lesbian, Gay, Bisexual, Transgender, Queer and/or Questioning, Intersex, and Asexual (LGBTQIA) colleagues, trainees and staff. We want to express our appreciation and respect for the important contributions of our LGBTQIA colleagues to our healthcare and science teams.

Some of our LGTBTQ+ colleagues have a fundamental question that should not even have to be asked any more in the 21st century: Where do I fit in? Like all of us, our LGTBTQ+ colleagues want to be accepted, included and validated. But unfortunately, many of them report experiences of hate, rejection and despair. This chapter will provide insights and ideas, how we can create a community where everyone feels valued and can make meaningful contributions. We want to overcome any kind of finite self-definition and get a more global perspective: We are interested in solving problems and helping other people. As Eleanor Roosevelt said, "Small minds discuss people, great minds discuss ideas."

So, what do we do when we see someone stuck in hate, anger and destruction? Instead of climbing into their dark world with them, putting all of us in it, we have the power to create a different reality. We can say: We see your small dark world, but our universe is different. Our world is abundant, colorful and beautiful. We cannot coerce inclusion and inspiration. But we can live it, and it will expand from there.

Dear LGBTQIA colleagues and trainees:
You are awesome and we know it!

How to be an ally

How can we support our LGBTQIA team members? One approach to find out is to ask them. Here are some suggestions that we learned through critical discussions. As with every chapter in this book, the points below are are not meant to be exclusive or comprehensive, but rather meant to spark ideas for tangible change.

- Publicly show your support with an ally symbol, such as a sticker, sign or poster. Add a rainbow poster at entrances, add a rainbow sign or sticker to checkin desks, public spaces, labs and offices. Wear a rainbow pin.
- Enforce equal employment opportunities and actively invite people from diverse backgrounds to apply for trainee, staff and faculty positions
- Challenge prejudiced and homophobic remarks and jokes.
- Be a role model for others by being open and visible in your support for LGTBQ community members
- Get comfortable with your own gender identity.
- Educate yourself on general principles of sexual orientation and gender identity.
- Become familiar with terms used by LGBTQ communities (see page 62)
- Create save spaces for LGTBQ+ personnel, including physical spaces where LGTBQ+ people and allies can meet each other and organization of conferences/events/ networking opportunities
- If you organize a diversity conference or event, formally include LGTBQ+ people
- Attend LGTBQ+ networking events. Be a visible ally.
- Invite LGTBQ+ trainees, staff and faculty to participate in and present at major internal administrative meetings and conferences
- Be a safe space: if someone needs to talk, be available, be there to listen.
- Offer training of personnel on policies and laws that affect gay, lesbian, bisexual and transgender individuals.
- Make sure appropriate HR procedures are in place to address harassment or discrimination against a person based on their gender identity or sexuality
- If you see inappropriate, bullying or harassing behavior against a member of the LGTBQ community, speak up. Silence communicates agreement.
- Correct myths or misperceptions, whether or not you know a LGTBQ+ person is present
- Healthcare workers: Politely and privately ask your LGTBQ+ patient, what name and pronoun you should use for medical documentations, legal and insurance purposes
- If you make a mistake, own it. Apologize.
- Provide single occupancy and gender-neutral bathrooms
- More information about how to create an inclusive healthcare environment for LGTBQ+ people: https://www.lgbtqiahealtheducation.org/wp-content/uploads/Ten-Things-Brief-Final-WEB.pdf
- Information booklet by the Joint Commission: https://www.jointcommission.org/-/media/tjc/documents/resources/patient-safety-topics/health-equity/lgbtfieldguide_web_linked_verpdf.pdf?db=web&hash=FD725DC02CFE6E4F21A35EBD839BBE97

Medicine is a very traditional and conservative field. Throughout my training, I have had the experience that some people in medicine do not want to adapt to the modern world. They feel offended by their gay colleagues and actively exclude them. It is like being back in high school: The easiest way for these bullies to bond is to talk smack about their gay colleague. The bullies that I had to endure said horrific, hateful things behind my back. They de-moralized and stressed me in many different ways: By shaming me with provocative gestures, by turning or walking away from a conversation, by exposing me in public, by blocking my access to important resources, by taking credit for my work, by humiliating me, by intentionally inflicting emotional distress - and the list goes on.

What happens in this completely choiceless situation has many parallels to physical assaults, including the fact that the victim is deliberately hurt, personally humiliated and systematically marginalized. People who experience these behaviors tend to withdraw, stop sharing ideas, stop asking for help to solve problems, and hold back. I spent days and nights trying not to cry. It is impossible to work productively in such a state of mind.

What can we do about this? The psychological problem with bullying is that the victim starts to internalize these hateful statements. When a bully told me publicly that I was "defective", I was terrified. What made all the difference was that one person stood up for me and asked the bully a simple question: What is wrong with you?

I literally came out strong: These experiences taught me what I stand for. I thank these bullies for showing me exactly who I do not want to be. I realized that I have the power to decide what to tolerate and what to focus on. There are many people who walked into my life and made it better. And there were others who walked out of my life and made it amazing!

Anonymous

"Some days there won't be a song in your heart. Sing anyway."

– Emory Austin

Working Toward an LGBTQ+ Friendly Stanford for All: The Stanford Medicine's Diversity Cabinet LGBTQ+ Sub-committee

Faced with the fact that significant LGBTQ+ efforts were lacking, Drs. Marcia Stefanick (Professor; Medicine, OB/GYN) and Jim Lock (Professor; Psychiatry and Behavioral Sciences) took on the enterprise of creating the LGBTQ+ Task Force within the Stanford Medicine's Diversity Cabinet. This task force is now considered the standing LGBTQ+ Sub-committee of the Diversity Cabinet.

Drs. Stefanick and Lock have recruited a dynamic group of representatives from different constituencies across Stanford Medicine as members of the LGBTQ+ Sub-committee. Sub-committee members represent the School of Medicine (SoM), Stanford Health Care (SHC), and the Lucile Packard Children's Hospital Stanford (LPCHS), and include faculty, staff, residents, medical students, graduate students, and post-docs.

The LGBTQ+ Sub-committee and its members continuously work on identifying needs of the LGBTQ+ community within Stanford Medicine, and then brainstorm, design, and execute plans to address these needs. As a result, numerous major ac-complishments have stemmed from the work of the Sub-committee members. Some of these include: 1) inception of the year-ly LGBTQ+ Forum event, 2) enhancement of the medical education curriculum, 3) LGBTQ+ faculty gatherings, 4) creation of an "out-list", through which individuals can self-identify as members of the LGBTQ+ community, 5) sensitivity training in the clinics, and 5) the first-ever presence of Stanford Medicine at the San Francisco Pride Parade.

But these efforts reach far beyond the Stanford Medicine community, and benefit the patients we serve. Members of the Sub-committee work on: 1) enhancing the LGBTQ+ patient experience, 2) developing healthcare programs for the LGBTQ+ community, and 3) educating the next generation of providers that will serve the LGBTQ+ population.

As concisely indicated by its vision: the LGBTQ+ Sub-committee strives to create an atmosphere that goes "from acceptance to belonging" for all members of the LGBTQ+ community here at Stanford Medicine. Actions give substance to words, and we - the members of the Stanford Medicine LGBTQ+ Subcommittee - are here to make it happen!

José Vilches-Moure, DVM, PhD, Dipl. ACVP

Assistant Professor
Veterinary Pathologist
Director, Animal Histology Services (AHS)
Chair of the Stanford Medicine Diversity Cabinet's LGBTQ+ Subcommittee
Stanford Medicine | Comparative Medicine

Proud at Stanford!

On the bright and sunny Sunday that was June 30th 2019, Stanford Medicine was represented in the San Francisco Pride Parade for the very first time!

Combined with Stanford University Pride, Stanford Medicine had a larger-than-expected presence. All units of Stanford Medicine were represented in this contingent: School of Medicine, Stanford Health Care, and the Lucile Packard Children's Hospital Stanford.

This vivacious crowd of LGBTQ+ folks and allies was composed of staff, medical students, faculty, physicians, res-idents, nurses, graduate students, post-docs, family members, and friends. As we gathered and marched, the laughter and smiles were plentiful. The celebratory mood was contagious. The camaraderie was unifying. The energy was palpable.

We hope this experience was as energizing and invigorating for the attendees as it was for the organizers, who look forward to many, many more Pride marches.

So be loud, and be Proud, Stanford Medicine!

The 2019 Stanford Medicine presence in the 2019 San Francisco Pride Parade was initiated and organized by Dr. Vinnie Alford (Post-doctoral Scholar, Stanford Institute for Stem Cell Biology and Regenerative Medicine) with the help of Robert Victor (Program Officer, SoM Office of Faculty Development and Diversity).

José Vilches-Moure, DVM, PhD, Dipl. ACVP

Assistant Professor
Veterinary Pathologist
Director, Animal Histology Services (AHS)
Chair of the Stanford Medicine Diversity Cabinet's LGBTQ+ Subcommittee
Stanford Medicine | Comparative Medicine

LGBTQ Workforce in Medicine: Mind the Gaps

This year, we recognized the 50th anniversary of the Stonewall uprising that launched the modern Lesbian, Gay, Bisexual, Transgender & Queer [LGBTQ] civil rights movement. Despite advancements in societal acceptance and legal protections, the size of the American LGBTQ population is still uncertain. Similarly, the numbers of LGBTQ health providers and their contemporary work experiences are also largely unknown. These data are important to obtain for many reasons. Firstly, there is still stigma and discrimination aimed at the LBGTQ population. As a result, there are observed health disparities. In crafting remedies for these disparities, it's important to understand how many people are affected. Further, in our culture, just the act of counting brings valuable visibility to policy makers and the general population. In order to deliver culturally competent care, our workforce needs to reflect the diversity of the communities we serve.

In 2019, the best estimate is approximately 4.5% [14.7 million] of Americans identify as LGBTQ. This percentage may seem small, but the number is significant. To provide a sense of scale, 14.7 million people is larger than the population of 46 of the 50 states. Further, these data only capture individuals who openly identify in surveys as LGBTQ. In essence, this is a measure of "outness" - a "floor" for the size of the LGBTQ population. It doesn't quantify the population whose concealed identities, attractions, behaviors and relationships don't reconcile with traditional models of heterosexuality or cis-gender identity. To gain any insights into the "ceiling" of the entire LGBTQ population, indirect means of measurement are needed.

A recent Stanford study reported while 80-85% lesbian and gay medical students openly-identify, only 45% of bisexual and 34% of transgender students openly-identify. The most common reason for not being out are, fear of discrimination, concern of career options and lack of support.

In October 2018, the Kaiser Family Foundation estimated there to be almost one-million physicians in the US. It is unclear if the prevalence of LGBTQ people in the overall population is reflected at the same rates within the medical profession. There are influences that could affect these numbers in either direction. Using the data from the Williams Institute and Gallup, it can be extrapolated that there could be between 37,000-45,000 LGBTQ physicians in the US. It's estimated there are between 38,000-47,000 radiologists in the US. Using the same logic as before, we can estimate there are 1,500-1,800 LGBTQ radiologists. Currently, 3132 physicians identify to CMS as Interventional Radiologists, suggesting there may be about 120 LGBTQ IRs.

In 2016, a study from Yale University School of Medicine, examined predictors for specialty choice for sexual and gender minority [SGM] students. Prestigious specialties, as measured by an objective index, are perceived by SGM to be less inclusive of SGM. The percentage of SGM in each specialty was

LGBTQ Workforce in Medicine: Mind the Gaps

inversely related to specialty prestige (P = 0.001) and positively related to perceived SGM inclusivity (P = 0.01). The "Most Welcoming" specialties were Psychiatry, Family Medicine, Pediatrics, Preventive Medicine and Internal Medicine/Pediatrics. The "Least Welcoming" specialties were Orthopedics, Neurosurgery, Thoracic Surgery, General Surgery and Colon & Rectal Surgery. It concluded that specialty prestige and perceived inclusivity predict SGM specialty choice. Further, SGM diversity initiatives in prestigious specialties may be particularly effective by addressing SGM inclusion directly. While it's unclear exactly how IR measures objectively in prestige, it has great similarities to the surgical disciplines that were "Least Welcoming". In the 2018 residency match, IR was by far the top of the list in competitiveness, inferring that it is in fact highly prestigious.

In a time when significantly more than half of medical school graduates are female, underrepresented minorities in medicine, and SGM, the "High Prestige" surgical training programs do not reflect the diversity of medical graduates. The sustainability of this dynamic is uncertain. However, it raises concerns regarding the ability to realize the full potential of the talent pipeline of graduating medical students in order to deliver culturally competent care to the population being served. As a strategic imperative for the future, the procedural disciplines such as IR must reflect on these dynamics in order to remain a vital link in the continuum of robust patient care.

Hirschel McGinnis, MD

Chair, Awareness and Collaboration
Steward Medical Group, Massachusetts
SIR Diversity and Inclusion Advisory Group

Glossary of LGBTQ+ terms for healthcare teams

- Birth Sex: determined by appearance of genitalia and chromosomes
- Cisgender: gender identity consistent with assigned birth sex
- Transgender: gender identity not consistent with assigned birth sex
- Gay: a person who is emotionally and sexually attracted to people of their own gender
- Gender: socially constructed characteristics of people as being feminine or masculine
- Gender identity: A person's internal sense of being man/male, woman/female, both or neither or another gender
- Gender queer: gender identity neither exclusively male nor female
- Sexual orientation: spectrum of attraction to others
- Queer: any identity not exclusively heterosexual
- Pansexual: attraction to many or all gender expressions
- MSM: men who have sex with men
- WSW: women who have sex with women
- FTM: female-to-male transgender individual (transgender male)
- MTF: male-to-female transgender individual (transgender female)

A more comprehensive list can be found under the link below. In this glossary by the National LGTBQ health education center, you will find some of the terms most relevant to the healthcare of LGBTBQ people. It is important to keep in mind: 1) Definitions vary across communities; not all of your LGBT patients will agree with all of these definitions; 2) There are many terms not included on this list; 3) Terms and definitions change frequently

Source: https://www.lgbtqiahealtheducation.org/wp-content/uploads/LGBT-Glossary_March2016.pdf

What Healthcare Workers Need to Know about Medical Imaging Studies of Transgender Patients

With transgender patients seeking care during various points in their transition, it is important for healthcare workers to familiarize themselves with appropriate medical terminology: The following publications provide a great overview and suggestions how healthcare workers can provide a welcoming environment and high quality care:

American Journal of Roentgenology. 2018;210: 1106-1110. 10.2214/AJR.17.18904
https://www.ajronline.org/doi/full/10.2214/AJR.17.18904?mobileUi=0

Health Services Research and Policy | Volume 18, ISSUE 3, P475-480, March 01, 2021
https://www.jacr.org/article/S1546-1440(20)31129-7/fulltext

"If the fire in your heart is strong enough, it will burn away any obstacles that come your way."

— Suzy Kassem

Gaslighting

"Gaslighting" describes abusive and dismissive behavior, specifically when an abuser manipulates information in such a way as to make a victim question his or her sanity. Gaslighting intentionally makes someone doubt their memories or perception of reality. Gaslighting in the workplace causes victims to question them-selves and their actions in a way that is detrimental to their mental health and their careers. They may be excluded, made the subject of gossip, they may be discredited or constantly questioned. The perpetrator(s) may systematically divert conversations to perceived faults or wrongs of their target. The result is constant undermining and destruction of the target's confidence, reputation and career progress.

Example: A colleague offers you a co-authorship in return for investing work in a project. Once you deliver the requested work, they provide you with a manuscript where your author position is different than previously discussed and documented. When you confront them, they turn the critique on you with comments such as "I don't remember we discussed authorships", "you are being so irrational" and "don't you think you are overreacting"? As a result, you question your own sanity and memory of the events. The gaslighter deflects from their own responsibility, making you feel crazy for even speaking up. By denying their promise, they get out of their obligation and discredit you in the process, so that all further critique of their behavior will be dismissed.

This list below demonstrates signs of gaslighting:

- You know something is wrong.
- You notice that negative gossip is being circulated about you.
- You face constant criticism.
- You ask yourself, "Am I too sensitive?"
- You feel gradually undermined.

WHAT TO DO

- Clarify ethical standards: For example, Stanford had a code of conduct, which clarifies what behavior is tolerated and what is not tolerated: https://adminguide.stanford.edu/chapter-1/subchapter-1/policy-1-1-1
- Don't confront a gaslighter directly: Gaslighters respond to criticism with personal attacks. If they feel threatened, they will retaliate and accuse you of wrongdoing in order to distract from their misconduct.
- Never be alone with a gaslighter: It will be harder for them to distort reality in the presence of witnesses.
- Offer to record meetings with the gaslighter: If the gaslighter opposes this request, document their answer. (Note that recording meetings in California requires consent by all parties. However, there are exceptions.)
- Document everything: Describe what happened, when it happened and how it impacted you and your institution. This will be helpful to validate your own emotions, understand the problem, find solutions and escalate to HR, if that is needed.
- In your log, always follow up with a note "I am valued", "I am worthy", "I am loved". This will counteract the negative impact on your mental health.
- Seek support from allies. You might be surprised by how many of your coworkers and leadership members will be ready to help and stand up for you.

Be yourself; everyone else is already taken.

— Oscar Wilde

Please remember, especially in these times of group-think and the right-on chorus, that no person is your friend (or kin) who demands your silence, or denies your right to grow and be perceived as fully blossomed as you were intended.

— Alice Walker
novelist, activist, and author of The Color Purple

The beauty of standing up for your rights is others see you standing and stand up as well.

— Cassandra Duffy

There's nothing wrong with you. There's a lot wrong with the world you live in.

— Chris Colfer

Being gay has given me a deeper understanding of what it means to be in the minority…it's made me more empathetic…it's also given me the skin of a rhinoceros, which comes in handy when you're the CEO of Apple.

— Tim Cook

You don't have to be gay to be a supporter. You just have to be human.

— Daniel Radcliffe

Notes

“ It's amazing how one day someone walks into your life, and you cannot remember how you ever lived without them.”

– Unknown

Marital Status

This chapter will reflect on how our marital status and family situation enriches and impacts our work. Our community consists of members who are single, living together, married, with or without children, close to or far from other family members, divorced or widowed. Each of these situations profoundly impact our work and our life.

Family members share joys and sorrows with each other. We celebrate successes and take care of each other when someone is sick or in need. The desire to belong to a community is wired into our DNA. Within this community, at home and at work, some people are deeply connected by a phenomenon, which we call "chemistry" or "connected souls". It is the most powerful force of human beings, stronger than the desire for money or fame and stronger than the decay or renewal of the human body. It can occur between two lovers, a parent and a child or two good friends. Connected souls can communicate without speaking, know without telling and provide comfort without the need for explanations.

We want to express our gratitude for the precious people in our lives who make our lives better. Please take a few minutes to be thankful, to be in the moment, and to appreciate that someone at work and at home. Do not wait to burn the candles, use the pretty tablecloth and open that special bottle of wine. Know that however good or bad the situation was today, it will change. And if today wasn't as good as you had hoped for, remember that most of us get a second chance: It is called tomorrow.

How to be an ally

For many working families, achieving work-life integration can be a struggle. This is especially true in the fast-paced field of medicine and sciences. Below are examples, how employees and colleagues can help each other during family transition phases. These points are not meant to be comprehensive, may not apply to everyone and may require tailored approaches:

- Know and clearly communicate rules around family leave
- Discuss with pregnant personnel if work accommodations are needed and desired
- Provide paid maternity leave and paid paternity leave
- Provide options for longer-term unpaid leave or part-time work arrangements
- Hire "floater" personnel or temporary personnel for covering longer term absences in order to minimize extra workloads and collateral stress for co-workers.
- Provide "on-ramping" services for personnel after longer-term absence
- Provide employee-assistance programs to help new parents find reputable child-care, a pediatrician and community groups
- Set up a lactation room. Provide appropriate time for lactation.
- Schedule meetings and conferences during working hours.
- Provide childcare options at conferences.

Every healthcare worker and scientist has to constantly balance work and life. Prolonged and excessive stress at work can be perpetuated by loneliness, increasing family responsibilities and/or other significant life events. If you notice physical, mental and/or emotional exhaustion and burnout of a team member, the following steps might be helpful:

- Burnout occurs in response to excessive workload. If you notice signs of burnout in a trainee, co-worker or yourself, decrease the workload.
- Set up realistic goals for yourself and others: Estimate the time and effort needed for specific tasks. Assign appropriate time for each task on a single calendar.
- Minimize interruptions during work hours as much as possible. Focus.
- Build a strong network of support personnel.
- Recognize signs of fatigue. Provide adequate recovery time.
- Schedule breaks during working hours. It is still common practice in the healthcare setting to assign employees to 8-10 hour working shifts without break.
- Eliminate mandatory conferences and meetings during "lunch breaks". A lunch break should be freely available to the employee.
- Schedule walking meetings for 1:1 or small group discussions
- Eliminate hidden energy-drains, such as after hour emails, online training courses and administrative requests. For example, if the institution requires an employee to complete 12 online training courses over the course of a year, schedule time for all of these during working hours. Eliminate unnecessary, redundant and/or inefficient requests.

"Love is composed of a single soul inhabiting two bodies."

– Aristoteles

"Some people make you laugh a little louder,
your smile a little brighter, and your life a little better."

– Anonymous

Our Shared Experiences

Marriage, like friendship, is about shared experiences. When I was looking for a life partner I had hoped to find someone who shared some of my passions and interests, like classical music and movies. The fact that I am married to a Radiologist is an extra special bonus because it lets me also share my work experiences and interests. As it happens, I met my wife Terry through our mutual involvement in contrast media research—we were actually "fixed-up" by colleagues who knew us both. Our relationship evolved in parallel with our research, and when I branched off into informatics, she understood my rationale completely and was supportive as I changed career directions. So she gets to use all the Radiology books I don't need anymore (!).

Terry Desser, MD, is Professor of Radiology and Associate Chair, Clinical Faculty Development

We are one of several Radiologist couples among the faculty, and luckier than most in that we are both at the same institution. When one of us has an especially good or bad day, the other often understands exactly why. But the downside is that we can sometimes spend too much time talking about work, and have to make a conscious effort to let things go once we are home.

Some of my colleagues have life partners whose schedules are completely flexible, which makes it easy for them to arrange travel and household tasks. That is definitely not the case for me, as Terry's Radiology schedule is dictated by clinical demands, so it is highly constrained. And this has only gotten worse recently as the hospital gets busier and busier. The logistics of my home life would definitely be easier if my wife had a job with a more flexible schedule. With a working spouse in a demanding field like Radiology, what is gained in extra household income is a trade-off with the added stress of trying to get all the necessary things done at home. But when I think of how much I value being able to share my work with Terry, and how much I gain by having a wife in Radiology to bounce ideas off of, I am sure it is worth it.

Daniel Rubin, MD, MS

Professor of Radiology, Biomedical Data Science, Medicine
Integrative Biomedical Imaging Informatics at Stanford
Stanford Medicine | Radiology

Finding the Right Job and Work-Life Balance

As the son of a urologist, it was often assumed that I would also become a surgeon. When I finished medical school, I was confronted with the many possibilities available in medicine.

This prompted questions that I was not ready to answer, such as what kind of lifestyle did I aspire to have? What were my priorities? What specialty could sustain my interest throughout a lifetime career? As a young doctor, I assumed it was possible to find a field which encompassed all of these priorities.

After much self-reflection, I discovered that surgery would not be a good fit for me, yet Nuclear Medicine was a perfect blend of all of my interests. It also allowed me to pursue a potential research and clinical future. Without the stress of overnight work or emergencies in the hospital, nuclear medicine gave me the opportunity to pursue medicine and spend time with my then two young children. We added another child and many other priorities to our list. I finished my training while my wife worked as a pediatrician, and once I became an attending, she decided to work part time.

When our middle son had a near fatal car accident, I was daunted by the fragility and finite nature of our life and made it my priority to ensure my career path was aligned with my initial career goals while constantly maximizing time with my family and attempting to be present during that time. At work, I became more involved in imaging informatics research, a passion left on hold during the time of clinical training and my initial years as an attending. At home, I made it a point to bike my children to school every morning to spend more quality, uninterrupted time with them. After dropping them at school, I bike to work which allows me to be physically active and adjust my mindset for the workday ahead.

Those precious moments between home life and work life have become extremely important to me as well- they provide the opportunity to cultivate activities (in my case biking, sailing, meditating) for myself that will further fuel and enrich my work and family life. These are only possible by having developed an intricate support system between my spouse and I, where we give value to family time, couple time and alone time. These are essential to maintaining some sort of work-life balance. We are fortunate to live in the warmth and beauty of the Bay Area which offers us a lifestyle that supports these values.

Guido A. Davidzon, MD, SM

Clinical Associate Professor of Radiology
Nuclear Medicine & Molecular Imaging
Stanford Medicine | Radiology

“No matter how far we come, our parents are always in us.”
— Brad Meltzer

Let's Reassess How We View, and Therefore Treat New Parents

The sum of our experiences, good and bad, makes us who we are, and affects how we think and approach problem solving at this moment. Every individual has unique thought patterns and strengths. When groups of people with varying thought patterns and practices come together to solve problems, we become stronger and more able to achieve the innovative and creative process desired to drive medicine and health forward. In this case, creating a culture of diversity is less about a moral responsibility and more about the best way to build a highly productive and innovative research environment.

One group of people, however, get the short end of the stick when it comes to being looked at for their abilities to contribute to a diverse thinking environment: young parents, especially mothers. Maybe it's the too frequent jokes about "mommy brain," or the fact that the bags under our eyes are a little larger, but in reality, as opposed to limiting or inhibiting our ability to contribute innovating thinking, becoming a parent can actually strengthen several of the skills that contribute to, and manage, diversity of thought. Parents of young children are forced to have excellent time management skills making them highly focused on the task at hand. They become more emotionally intelligent, better able to read group dynamics and interpersonal relationships. Becoming a parent can tighten the drive and motivation behind pursuing research as we're forced to evaluate our priorities in life, and can even make us more ambitious as we strive to become role models and provide for our children. We actually become more courageous. Biologically, the increasing oxytocin levels in our systems, both moms and dads, decrease activity in the brain from fear stimuli, perhaps leaving us more able to express our thoughts and opinions. Perhaps it's time to stop only tolerating the presence of young parents in our work environments, overlooking our new experience and abilities. Instead, view us as what we are: finely tuned and highly optimized individuals ready and motivated to innovate and excel in group dynamics, not in spite of, but because of, our role as parents.

The academic leaky pipeline, or the slow trickle of women out of academics with the progression of career level, becomes more of a deluge when female academics hit the child rearing years, coincidentally just around postdoc or junior faculty levels. This is a lose-lose situation. Institutions end up lacking diversity of thought and women miss out on good jobs. Many ideas have been proposed to help this problem at the institutional level, but we all know these large, system level advancements take time. Perhaps, an immediate thing each of us can all do is reassess how we view, and therefore treat, new parents in the academic research environment. We should highlight what skills and advantages they bring rather than operate from a deficit perspective centered on the perceived loss of continuous availability and, in essence, make sure our ideas pivot from the archaic stereotypes. Now that would be a diverse thought.

Katie Wilson, PhD

Instructor of Radiology
Stanford Medicine | Radiology

Family Support Creates Peace of Mind

I had the privilege of joining the Department of Radiology as an Assistant Professor in November 2015. I am married to a wonderful man named Alex and we have a lovely baby girl, Elena. I wanted to share my story with you as this Chapter is dedicated to marital status and how it enriches our work.

My story began in 2011 when I met my husband, Alex, in Los Angeles while doing a postdoctoral fellowship at UCLA. Alex is an automotive designer and in early 2013 he got an outstanding career opportunity at Chrysler which required him to move to Michigan. While I stayed back in Los Angeles to complete my postdoctoral fellowship, Alex flew from Detroit to Los Angeles three times a month for over two years. In late 2014, he started looking for a new career opportunity and I began my faculty position search, hoping that Alex and I would finally get to live together. A week after Alex got an offer from Volkswagen in Santa Monica, I got an offer from Dr. Gambhir to join the Department of Radiology. I still remember calling Alex that day, telling him the great news with little sadness in my voice as it meant being apart once again. He congratulated me and said "This is a fantastic opportunity and you will do great. This is your dream and you should go after it. Regarding us, we will be in the same State—we are getting closer and closer….and the flights are getting shorter and shorter….it will all work out and will be great, you will see". His response was so positive and inspiring that gave me great confidence and excitement about my move to Stanford. I even started thinking about the projects and proposals I would like to write for my lab that very same day. Soon after, Alex moved to Santa Monica, we got married and a month later I moved to Stanford. I had already prepared myself for at least a few years of long-distance marriage or as my husband liked to say, "short distance—at the end we were in the same State". A few months after joining Stanford, Volkswagen decided to relocate their studio from Santa Monica to Belmont. What were the odds? Never in a million years did we think that would happen for us. I like to think of it as fate.

We now have a lovely daughter, Elena, and she means the world to us. In addition to my husband, I have tremen-dous support from my mother, Elena, who moved to the US to help us care for our baby girl. Having our family recognize and support the passion for our work is extremely valuable and I truly believe it enriches everything we do in science or any kind of work. I am never worried when I need to work late or over holidays as I know that I have the support of my whole family. This piece of mind helps us focus, be efficient and move forward. Addition-ally, having an artist right beside me in science definitely helps me get our work featured on a cover of a journal. I do consider myself extremely lucky having the opportunity to be here and be surrounded and supported by all of you and my family!

Tanya Stoyanova, PhD

Assistant Professor of Radiology
Canary Center at Stanford for Cancer Early Detection
Stanford Medicine | Radiology

Above: Spring 2018, with my children in Patagonia, Argentina. I feel lucky that they share with me the love for travel, adventures and the mountains.

At left: My son at the age of 11. Both of my children helped me to enrich my collection of teaching cases of buckle fractures.

My Thoughts on Marital Status... as a Single Parent

I was asked to write my thoughts on marital status... as a single parent, a pediatric radiologist and as an accomplished rock climber. It was very nice to be seen this way because, most of the time, I felt my life and career were just a tentative way to keep my head out of the water. I went through a painful divorce eight years ago and since then have been raising my children alone. Well, not quite alone, I had support from other radiologists, covering for me in times I had to go to court for my divorce, when I had to pick up my sick children at school, meet the teachers or go watch my kids in a school show. My parents have been also absolutely wonderful, flying from Europe to help me at home when I had to go to meetings or when I was totally exhausted.

My days start way before I go to work, and end late at night, when dinner is cooked, children are in bed, laundry is done and lectures or papers ready. Although alone, I wanted to fulfill all my role as a mother and I didn't have a nanny. I cook dinner every night, do all the laundry, and used to tell stories at night when my kids were little. Now that they are teenagers, I am trying to morally support them. I never planned to have a big career but I really enjoy my work. I like to take care of patients, I like to teach and I always loved to do research. Being a single parent, I have learned that I need to do things not only well, but also fast if I want to be successful. I have also learned to prioritize, whether it is to talk to a family of a sick child, dictate and sign my reports, find the time to fill up the fridge, or pick up my kids on the street when they fall from a bike. I have learned to improvise as well. I remember taking my four and six-year-old in the middle of the night to the hospital to reduce an intussusception, asking the parents of the sick child to keep an eye on my own children. Being a mother and single parent definitely made me more sensitive and companionate to the stress and worries of other parents and their children. I also believe it is important to do different things additionally to my work. For me it's also important to travel, to be interested in art and literature, to discover other cultures, to keep my mind open and to keep my daily work in perspective.

Erika Rubesova, MD

Clinical Professor of Radiology
Pediatric Radiology
Stanford Medicine | Radiology

Being a Widow

It is very difficult to properly describe the pain and grief we have been through since January 8, 2018. That day we woke up like any regular Monday morning, unaware that life, as we knew it, would be over just a few hours later. Within a split second our lives changed from hectic, but extremely happy and pretty much picture perfect to complete shatters. Not a single day goes by without me catching myself wishing it was all just a nightmare, that I can wake up from it and all is back to normal again. The children often speak about wishing for a time machine to revert the tragic events. We just miss him so very much.

The implications of Juergen's death are quite frankly suffocating. As if it were not already enough to lose the love of your life or your great father, respectively, we were drowned in an avalanche of problems, ranging from financial aspects, the house, career implications, emotional harm, truly existential problems overall. Whenever I feel that we manage somewhat, a new issue arises. I always had the deepest sympathy and respect for single mothers, but that was from the safe distance of being in a stable relationship where both partners pulled their weight. While I certainly had Juergen's back so he could advance his career in record time, he truly cared and was deeply involved in all things related to our family. Family was one of his top priorities. I cannot possibly fill that void. Consequently, the children have been very traumatized. Juliana, only five years old at the time, still deeply affected, is making some progress. But her older brother Alex suffers deeply. He had such a strong bond to his father and vice versa. While he slowly starts talking about dad, he is often still in denial and tells himself that dad is in China for his research and will come back some day or some other halfway plausible scenario.

I do not know what would have happened to us without the great support that we received in the months following Juergen's death. Sam and Aruna who were pitching in from the very first hour after the accident, so many of my colleagues, friends and neighbors helping out with dinners or offering play dates and welcome distraction for the children, lending an ear and some perspective, my colleagues in the musculoskeletal imaging division who are so supportive and understanding, and our nanny, Elke, really more a family member, who is so committed and one of the most important constants in our lives. We are deeply grateful for the enormous compassion and generosity.

Now, more than 9 months out, life is unfortunately still a daily struggle. We have found some new routines, but nothing really feels normal yet. I should be completely focused on my career, which is very challenging in the current situation.

There are definitely days where I feel that I can tackle this, but also still so many days where I am just completely paralyzed, and grief struck. I refuse to be a victim of our circumstances, but can only hope that I will grow with my challenges. We have a long way to go, it will take time, and we will need a lot of help along the way. To me that is actually one of the most difficult aspects: asking for help.

Amelie Lutz, MD

Assistant Professor
Co-Division Chief, Musculoskeletal Imaging
Stanford Medicine | Radiology

PROFESSIONAL FULFILLMENT AND BURNOUT

Professional fulfillment is defined by terms such as work satisfaction/happiness, meaningfulness and self worth. Contrarily burnout is characterized by work exhaustion and interpersonal disengagement.

Which of these sound most familiar? For myself as well as others, both apply at different times in our lives and careers, sometimes both in the same week. Although radiologists have interesting, intellectually challenging and varied professional lives, we are unfortunately also often stressed and increasingly isolated at work in this busy computerized age. Physicians in general have high rates of burnout and radiologists are similarly affected. So how can we tip the balance more often towards professional fulfillment and thereby decrease burnout?

The seven drivers of professional fulfillment apply to radiology as much as to other professional fields. These include workload, efficiency, control/flexibility, values alignment, collegiality/community at work, work life integration and meaning in work.

Based on my own experiences in pediatric radiology as well as conversations with others, this is a commentary of how some of these may apply to Radiology in general and Stanford in particular.

1. **Workload** – this has increased exponentially over the past few years, throughout all our divisions. Staffing has not kept up with rapidly increasing clinical demand and apparent urgency, complexity of cases and technology. There is less time available for meaningful interaction with patients, clinicians and colleagues as we struggle to get through the overwhelming volume of work. The academic focus of our department achieves amazing results but also removes faculty from the clinical sphere, placing greatly increasing strain on those who are more avail-able. Working with trainees is a double-edged sword, often helping enormously to complete the work but also adding to the complexity of tasks because of the time and effort involved in providing supervision and teaching.

2. **Efficiency**- " A workman is only as good as his tools." The main ancillary tools to support our own knowledge, experience and academic resources are equipment, technologists, PACS, administrative and support staff. All of these have become increasingly challenging. Recruiting and retaining adequate numbers of high quality, well-trained technologists and administrative staff is difficult in an expensive living environment such as the bay area. Radiology PACS inefficiencies, slowdowns and breakdowns can also be a significant stress factor for everyone. A new PACS system can provide some improvement but may not be a quantum leap, time will tell.

3. **Control/flexibility** – does it feel like we have very little? Radiology provides important consultation and imaging services to many different clinical areas. The service aspect tends to be over emphasized with seemingly every-one's needs except our own taken into account. This can result in very long workdays and poor work-life balance. While there are some mitigating factors including our hardworking trainees and use of a teleradiology service, some specialty areas such as cardiac imaging and complex MR/CT services are sometimes not covered. On the plus side, people do help one another when possible and there is a good sense of community and support when someone is ill or otherwise indisposed/unavailable.

4. **Values alignment-** The radiology leadership and faculty goals appear generally well aligned. We all strive for a supportive, appreciative environment while providing high quality, caring, timely, innovative imaging services and cutting edge research. With translation into practice there appears to be less good alignment in such areas as keeping up with staffing needs relative to increased volume, coverage of multiple services simultaneously and support for reasonable life-work balance.

5. **Meaning in work** – we all like to think that our work is meaningful and important and strive to make useful contributions to patient care. While imaging has become ever more diagnostically important in this technological age, our efforts and contributions may not be appropriately recognized or appreciated by our clinical colleagues. There are still half joking comments about the short hours that radiologists work (wish that were true), while it only takes a brief compliment or acknowledgement to make a big difference to someones' day.

I am currently the appointed wellness director for the department of Radiology at Stanford. I have spent many months trying to read and educate myself about the issues surrounding physician wellness and burnout. I have spoken to a variety of other faculty; attended wellness MD information sessions; heard about problems and programs in other departments; connected with overlapping department initiatives such as diversity and professional improvement and discussed how to impact physician wellness in radiology in our many different divisions with some common but also many different challenges. Suggestions and programs in other departments have included commit-tees or groups looking at work distribution and scheduling improvements, focus/brainstorming groups, gender/diversity evaluation, mentoring relationships, sponsored social groups e.g. for younger physicians or faculty with similar interests/issues, meditation/massage/exercise/self help sessions, ergonomic evaluation, intervention and counseling resources.

I would be happy to connect with others regarding their thoughts and ideas on physician wellness and professional fulfillment in our department and what our priorities should be to work together to improve the Stanford radiology community. There is not a universal miracle solution to burnout, but probably small improvements that can gather steam. Stanford is generally striving to be a leader and example to other institutions in regard to physician wellness. Our leadership is similarly demonstrating a real commitment to this area by announcing the establishment of a new department vice chair position for physician mentoring and wellness.

Beverley Newman, MD

Associate Chief Pediatric Radiology
Stanford Radiology Wellness Representative
Stanford Medicine | Radiology

An Interview with Dr. Tait Shanafelt

Q: What is your role at Stanford and what are the goals of your office?

A: As Chief Wellness Officer and Associate Dean, I help direct the WellMD/WellPhD Center. Our goal is to organize and help develop the strategy for advancing professional fulfillment and well-being for physicians and scientists in the School of Medicine and to help catalyze improvement at the level of departments, divisions, and the hospitals to make progress. The role of our team is to help measure where we are, provide honest appraisal of that to our senior leaders, and to help advocate for the well-being of physicians and scientists with school and hospital leaders as decisions are made. We provide data to the department chairs and divisional leaders to help guide and inform the department-specific actions. We also provide support and improvement tactics to those leaders depending on what targets they prioritize as the biggest issue in the department so that they can hopefully be effective in driving positive change. Often there are things departments can implement within the next 2-3 months that make people's lives better.

Q: Do you have an example?

A: Improving the scheduling system, equity and transparency around a whole host of characteristics (pay, night, and weekend shifts), optimizing cross-coverage systems, insuring that the electronic resources being used are operating as they should, improving work-flows, enhancing community and reducing isolation. The opportunity areas are often distinct for each specialty. When the focus is on generic things designed to benefit the whole school or medical center, we often miss the local issues that matter most. What people really want is for leaders to address the friction points in the daily work, not the things that I have to deal with once in a while. If local teamwork is sub-optimal, if the tools I have to use are regularly breaking down, if inefficiency in the work flow is making me go home late or perform excessive work at home and that's eroding my relationships, those are the things that really grind people down.

We often simplify the improvement dimensions into seven domains: workload, efficiency, flexibility and control over work, work-life integration, improving meaning in work, collegiality and community among colleagues, and culture and alignment of the values in the department. Usually, those are typically the opportunity areas, which of those is at the top priority, how it is manifest, and what change would be helpful varies by department.

Q: If you have administrators in a department that start to see signs of burnout in their employees, what are some of the things that you would recommend they do?

A: First, we need to engage as a leadership team and acknowledge that we see the problem and believe it's important. Until we prioritize it as a leadership team, we often come up with over-simplified quick-fixes that put the blame on the individual to take better care of themselves - sleep, self-care, nutrition, exercise, mindfulness. There's nothing wrong with any of those things, but in a sense, what we're telling people is to become more resilient so that you can tolerate a broken work environment, instead of focusing on fixing the broken work environment that's causing the problem.

The first step is to listen to our people. One of the mistakes that we often make as leaders is that, once we recognize, this is a real issue and it's having important consequences, we close the door with the leadership team and ask "what do we need to do to fix this"? Oftentimes, our perception of what people want fixed, what those local broken windows are, is inaccurate. Recently, we gave every department chair the report for their physician-wellness surveys. We also had free text comments asking physicians "If there were one thing your department could do to improve the well-being of the physicians in the department, what would it be?". I sent the results of the free text comments to the chair of a large department this afternoon and they responded within 15 minutes shocked because the comments were not at all what they had expected people wanted them to work on.

Q: If you are on the other end of the power spectrum – trainees, residents – and you see inefficiencies, who do you talk to or where do you start?

A: First, if the individual is in distress or at a crisis point they need to get help. Things might have been okay a month or two ago but suddenly they workload has changed or something else has occurred in their personal life - their partner loses a job or one of their parents is sick, and all of a sudden a work load that was okay last week is now sinking me. So, how do we create low barrier, low stigma resources for those times of need? In addition to the university help desk, there are a number of specific resources for physicians and residents/fellows. The WellConnect program for residents that is a team of psychiatrists on call every day (https://med.stanford.edu/psychiatry/special-initiatives/wellconnect.html). We also have the peer support program for physicians (PRN Support: http://wellmd.stanford.edu/get-help/prn-support.html). If an individual is struggling, then let's get them individual help.

If the issue is a broader one that residents see a need or opportunity to improve our system, they would ideally be working with their program director. In an ideal world, the program directors create regular forums to discuss such opportunities just like we're hopefully doing for faculty.

Q: Is there anything else you would like to add?

A: Diversity, inclusion, and wellness are inseparable. People need to feel like they can be their authentic self, that they're treated fairly, that who they are is valued and appreciated, that they have colleagues that support one another – those things are foundational. If we are not improving those characteristics, it breeds a lot of other discontent. Attending to diversity and inclusion is something we must improve in its own right, but if we're not doing that effectively, it's going to undermine all the things we're doing to promote wellness and professional fulfillment at the same time. They are interconnected.

Dr. Shanafelt was interviewed by

Jessica Klockow, PhD
Chair, Radiology Trainee Diversity Committee
Postdoctoral Fellow
Stanford Medicine | Radiation Oncology

Notes

MEN IN STEM

"Love isn't a state of perfect caring. It is an active noun like struggle. To love someone is to strive to accept that person exactly the way he or she is, right here and now."

– Mr. Rogers

Men in STEM

This chapter is dedicated to our male colleagues, trainees and staff. We appreciate your devotion and important contributions to the field of science, technology, engineering, mathematics and medicine (STEM). We want to send a specific note of appreciation to our male colleagues who support their female team members and team members from underrepresented minority backgrounds. Your voice, your insights and your vision are amazing!

A big part of creating an inclusive culture is about increasing awareness about the impact of stereotyping and its effect on confidence and the ability to contribute. If you are a white male, you likely enjoy white privileges. Biases do not target you personally most of the time. If you have never been part of a minority group, you might not fully grasp the cruelty of constant criticism, disapproval and rejection – and its impact on our collective morale and productivity. If you are reading this book, you decided to educate yourself beyond your current fund of knowledge. On behalf of our community, I want to thank you for taking the initiative! You our the most powerful ally!

Some men in our community are also members of minority groups, because of their race, religion, gender identity or other characteristics. They are sharing some of the biases and obstacles they are facing on a daily basis. I want to thank you for speaking up! Your strength and perseverance is inspiring!

Why should the lucky members of a majority group care about diversity? Firstly, we are all members of an increasingly complex world: We cannot be at the top of the innovation curve if we do not form teams with a variety of skills and abilities. We need disruptive ideas, diverse insights, multi-lingual competencies and different viewpoints to solve increasingly complex problems. In addition, demographic, cultural and political power structures will change. Anyone of us might find themselves to be a minority in some place and at some time. If we created a world where the term "minority" has been replaced by "rare and unique", this will not have negative impacts any more. Because we created a meritocracy where the best ideas will win!

Men in STEM can be members of minority groups and be targets of biases, discrimination and violence. Specific action points and interventions may vary depending on the specific minority group they associate with. We refer to the respective chapters for examples of allyships for these team members. The points below provide examples, how men - and white men specifically - can serve as powerful allies for team members from underrepresented minority (URM) backgrounds. These examples are not meant to be comprehensive and may require adjustments for specific situations:

- Continuously educate yourself and others about diversity and inclusion matters
- Recognize that you have unearned privileges that ease your path to success
- Recognize that there is more than one experience
- Do not dismiss someone else's concerns because you experienced some hardship as well
- Do not make this discussion about you. Focus on the other person. Listen.
- If you find yourself in a privileged position of power and/or wealth, find a way to give back to the community in a way that makes sense to you. Be a role model and inspire your friends and colleagues to give back to their communities as well.
- Do not try to rescue, mansplain or become a spokesperson for minoritized people
- Use your position of power to effectively confront another man's bias
- Promote products and ideas of women, people of color and other URM members
- Use your privilege to advocate for URM candidates for appointments and promotions
- At meetings or conferences, do not dominate the discussion. Create new approaches to collect as many viewpoints and ideas as possible.
- Attend to group dynamics to ensure the inclusion of women and people of color.
- Communicate information from the dominant group to URM team members
- Invest your time and resources to serve as mentor and sponsor of URM team members
- Examine your meetings and conferences for female and URM representation
- Use your privilege to invite female and URM team members to attend meetings where corporate decisions are being made
- Invite your white male friends to participate in diversity and inclusion initiatives: Give your white male friends a list of tasks and most of them will help
- Use your privilege to introduce members of minority groups to your powerful friends
- Speak up when you witness problematic statements and/or behaviors
- Challenge the status quo in the classroom through the topics discussed, new teaching styles and active solicitation of different viewpoints
- Avoid sharing content that is traumatic
- If you have managerial authority, use it to shape policies for appointments, work assignments, teamwork, leadership development and promotions
- Humbly and actively implement anti-racist policies
- Create a group or committee of allies, who can improve diversity and inclusion at your institution from different angles
- Introduce incentives for effort on diversity and inclusion initiatives
- Develop strategies to successfully engage points of view that you find unattractive
- Critically discuss bias, discrimination and structural racism with your friends and colleagues from majority groups who are skeptical about white privilege
- You do not have to be black to support Black Lives Matter , do not have to be a woman to support the #MeToo movement and you do not have to be gay (or member of the LGBTQ+ community) to attend a pride parade.

Everyone has a Story

I came to Stanford 23 years ago, as a graduate student in electrical engineering. (Yes, do the math, that means I was just 8 years old!!). I had been working in a company that I loved, but frankly was tired of working on version one of a product when version three was being planned already - the idea of developing a product just to move money around, and which would be obsolete in a year got old. So I came to Stanford grad school, got my PhD, and ultimately found a position in radiology.

Along the way I've been fortunate with opportunities, and been impressed at the efforts Stanford often goes to in embracing diversity. As a research associate in graduate residence I met some 200 new people each year, in all disciplines. Before long I would live the principle of "What's your story?" Everyone had some unique experience, from moving across the world and learning a new language and culture, to moving across the bay and rooting for the wrong college team. The key was to talk to people and get to know those experiences, because every single person had something interesting to share, if you could just get to talking about it. I feel very fortunate in Grad school to have interacted closely with the most diverse group of peers than at any point in my (sheltered) life.

Now things are a bit busier. The information firehose never ends for most of us. I brew my coffee while I fetch the cream so as to get back to work as fast as possible. But everyone still has a story, and it's just a matter of talking about it. Oddly most people are comfortable sharing something personal that, if you take the time to stop and think, challenges your views that have often been simplified to maximize efficiency. So my challenge, to you and to myself, is to take some more time to talk to the people around you. Talk to staff, students, faculty and postdocs in your workplace, talk to mothers, fathers, aunts, uncles, brothers, sisters, MDs, PhDs, BAs, from all over the world. They've all had some experience, and probably have something really important to them that most of us never know. So go and find out a unique fact about that person you see most days, and you'll be better for it.

Often we talk about 'science communication' and how we must learn to tell our story. But equally important is to hear the stories from others. We live in a bubble - if we can't learn from the diverse experiences of people at in our community, how are we to go out into the world and help solve much greater challenges, with even greater diversity of cultures, wealth, abilities and much more? I look forward to hearing your stories, and those you hear!

Brian Hargreaves, PhD

Vice Chair for Research
Department of Radiology
Stanford University

Dr. Larry Chow is a busy radiologist and also a hobby photographer with more than 6,000 followers on instagram. A selection of his beautiful photos is shown above.
https://www.instagram.com/batmobile88/

Lawrence Chow, MD

Director, Emergency Radiology, Body Imaging Division
Stanford Medicine | Radiology

Leading Global Health Efforts

I am excited for the opportunity to represent clinical imaging with Jayne Seekins as part of the Stanford Center for Innovation in Global Health (CIGH) PLC (power line communication) network (https://globalhealth.stanford.edu/about.html). My background in global health dates back to 1999 when I lived and worked in Kosovar refugee camps and continued during my medical training as I served as vice president at Rad-Aid (https://www.rad-aid.org). Rad-Aid is the largest international radiology global health non-profit organization. I oversaw the establishment of outreach (clinical and education) programs for Rad-Aid in China and India and created the largest international resident trainee chapter program, which has since grown substantially and now includes representatives from more than 30 academic institutions serving 14 countries. It has also been an honor to once again serve as senior editor on the first textbook on the subject of imaging applications in global health titled "Radiology in Global Health", which is now in the second edition and which we have just finished for release later this year. Finally, my current research efforts using artificial intelligence applications in medical imaging have led to an exciting collaboration with Jayne Seekins where we are leveraging AI models on a smartphone platform to provide instant point-of-care world-class radiology diagnostics to parts of the world that lack radiology expertise - our hope is to find partnerships through our work on the committee here at Stanford to expand this project. We are hopeful that these new technologies could provide a powerful supplement to resource-poor populations all over the world for sustainable screening and diagnostic applications.

Matthew Lungren, MD, MPH

Associate Director, Stanford Center for Artificial Intelligence in Medicine and Imaging
Stanford Child Health Research Institute Faculty Scholar
Associate Professor of Radiology, Department of Radiology
Stanford University School of Medicine and Lucile Packard Children's Hospital

Diversity is about Empowering Individuals

The April 2017 feature article of The Atlantic magazine was titled, " Why is Silicon Valley so Awful to Women." (https://www.theatlantic.com/magazine/archive/2017/04/why-is-silicon-valley-so-awful-to-women/517788/) .

It was disheartening to think that the epicenter of innovation and progressive societal disruption could foster a culture of exclusivity and inequality when it came to gender. It was even more disheartening to see the comments from some physician colleagues on the social media platform Doximity in response to Dr. Daldrup-Link's efforts to promote diversity in our Radiology Department. They argued that a Radiology Department should not hire diverse team members, but the best and the brightest. They apparently did not realize that diverse team members could be the best and the brightest.

I personally think these physicians are missing the point. Promoting diversity and inclusion in our field is much bigger than setting quotas for minorities and women or setting up "reverse discrimination" policies to lower standards by which we recruit physicians into residencies and practices. This is about empowering individuals that are overtly or passively discriminated against and ensuring they have the opportunity to prove that they are able to become professionals. Wage disparity is a fact in all industries so the claims of rewarding people in a meritocracy do not always hold muster.

We also owe it to our patients to diversify our faculty to reflect the communities we strive to treat. We all must do our part to promote the diversity of our faculty and staff. It will take significant struggle and sacrifice. However, in the end, these efforts will make us a better health provider, colleague, department and institution.

Safwan S. Halabi, MD

Clinical Associate Professor, Department of Radiology
Stanford University School of Medicine

A Few Questions About Microaggressions

The primary difficulty with addressing microaggressions is the term itself, which implies brief incidents with minimal consequence. In his best-selling 2015 book "Between the World and Me"[1], Ta-Nehisi Coates presents a letter to his adolescent son explaining how to find his place in society. He discusses how America was built on the idea of "race", a concept which was deliberately weaponized as a means of maintaining social order. One of the most effective components of this arsenal is microaggression.

Microaggressions are often delivered via seemingly benign or even complimentary spoken words. A 2007 article in American Psychologist by Derald Wing Sue[2] lists a variety of examples of racial microaggressions encountered in everyday life. Invariably they all reinforce an underlying narrative that the recipient is somehow inferior. One example Sue provides is telling a person of color how articulate they are, which implies that it is unusual for someone of that race to be articulate. Another example is promoting the myth of colorblindness through statements such as "When I look at you, I don't see color" and "There is only one race, the human race." Even though Sue's piece was published over a decade ago, its current relevance is easy to recognize in the misguided phrase "All lives matter". Choosing not to acknowledge someone's race, when relevant, is similar to ignoring the fact that someone is 7 feet tall or using a wheelchair. While these differences should not affect the overall quality of their work, they will affect how they approach their job. Some tasks will be easier, others will be harder, and many will just be different. It is imperative that healthcare providers embrace these differences in order to recruit the most talented individuals and provide the best care for our patients.

In healthcare settings, which are often a high stress/high stakes environment, microaggressions can have an effect which is far more substantial than intended. Let us all take a moment to reflect on our experiences with microaggression in the workplace. Have you heard any of the phrases in Sue's paper[2] used in the hospital? Have you ever heard anyone tell a physician/nurse that they speak English well or ask where they were born? Were such comments ever directed towards you? If so, how did it make you feel? If not, how did it make you feel? Did you say something about it? Do you think you should have? Knowing what you know now, what will you do next time?

Kevin C. McGill, MD, MPH

Assistant Professor
Musculoskeletal Radiology
Department of Radiology and Biomedical Imaging
University of California, San Francisco

REFERENCES

[1] Coates T. Between The World And Me. Spiegel & Grau; 2015.
[2] Sue DW, Capodilupo CM, Torino GC, Bucceri JM, Holder AM, Nadal KL, et al. Racial microaggressions in everyday life: implications for clinical practice. Am Psychol 2007;62(4):271-86.

Examples of Microaggressions

Microaggressions are subtle verbal and non-verbal insults, often done unconsciously. They are layered insults based on one's race, gender, class, sexuality, immigration status, accent, or surname. Examples:

- Introducing the female doctor by first name and the male doctor by Dr. Last name
- Cutting a Black team member short when they communicate
- Being told "we don't have racism here" by an all white/asian leadership
- A Black faculty receiving zero mentorship requests from white or Asian students
- Seeing people asking my junior white student for advice rather than me
- That question: "Is this your real hair"?
- Receiving the feedback that I am quite assertive in meetings
- That comment: "Oh, you are a real doctor"?
- Constant advice - what I should be doing differently
- Absence of entirely positive feedback
- Every praise is followed by criticism
- Assumed to be guilty if anything negative comes up
- If there is a dispute, I am asked to apologize without anyone asking me what happened
- Having to research how a county treats black people before planning a conference trip

What Allies Can Do Instead:

Examples of Micro-inclusion

- Listen
- Believe others' experiences. Do not assume that something cannot happen simply because you did not personally experience it
- Address MD and PhD minority members with their title
- Invite minority members to share their opinions
- Ask minority members for advice
- Thank minority members for comments or actions that were helpful to the team
- Acknowledge a great idea
- At a meeting, repeat a comment and provide credit: I agree with Lisa's suggestion to improve.
- Normalize changing your opinion when new information is presented
- Introduce minority team members to sponsors and influencers
- Praise without attached criticism
- Advocate for inclusion of minority members at important events
- Always speak up if you witness hatefull or ignorant comments
- Provide leadership opportunities for minority members

Anonymous, Stanford Radiology Diversity Committee

Diversity Matters - No One Should Feel Shut Out because of Who they Are

It is no secret that historically, radiology has been, and continues to be a male dominated specialty. Women make up about 46-47% of all medical students, yet radiology residencies comprise only about 27% women. Interestingly, in 2016 the AMA placed radiology as number 9 in the top 10 specialties for women based on work-life balance, predictable and reasonable work hours, compensation, and significant impact on patient care. Despite this endorsement, the number of women entering radiology residency programs, unfortunately has not changed significantly in the past few years.

This begs the question, why aren't more women entering the field of radiology? The answer is multifaceted, but data from a 2013 study from the Journal of the American College of Surgery found that women are more likely to enter specialties with higher proportions of female residents, as these women directly serve as role models to women medical students. Furthermore, certain specialties tend to offer more in the way of patient contact, continuity of care, and long-term patient relationships that may appeal more to some women. There also may be a perceived sense of work-life imbalance with radiology, or conversely, a lack of knowledge of what radiology actually does offer in terms of work-life balance. Radiation exposure and its effect on fertility has also been raised as a concern for some women. Lastly, one cannot discount how the culture or perception of a specialty's culture influences the type of people attracted to said specialty. From the outside looking in, radiology has all the makings of a Boy's Club, so those who "fit" the mold tend to usher themselves into the party, and those who do not look elsewhere.

But why do we care if women aren't entering the field of radiology? Because diversity matters. Not only gender diversity, but also racial, ethnic, and religious diversity. Our work force needs to reflect the patient populations we serve because perspective matters. The perspective offered by having a diverse workforce made up of people that can relate to our patients from different backgrounds cannot be understated. As men who dominate the field, I believe it is our duty to strive for inclusion and our charge is to do any and everything we can to increase the number of women entering into the specialty.

Again, for many years radiology has been touted as a great choice for women, yet the number of women entering the field remains stagnant. So how do we tackle what seems likes such an insurmountable obstacle? Certainly, we could come up with a laundry list of very reasonable tactics and strategies, but if we could start small and focus on 3 principle areas, our efforts would go a long way.

Exposure – Early exposure to radiology is critical. Radiology is not a required medical school clerkship at most medical schools across the country. In fact, many of us currently in the field could attest to the fact that we did not discover radiology until late in medical school, or stumbled upon the field late by complete chance, or never really even knew what a radiologist did down there in that dark room until we began pursuing the specialty. Further, how many of us recall receiving any lectures from radiologists during the 1st and 2nd years of medical school? Or how many of us recall having a radiology resident, fellow, or attending as a gross anatomy lab instructor during our 1st year of medical? I certainly did not experience any of the above, and who better to teach anatomy than a radiologist? As radiologists we need to more actively insert ourselves into the medical school curriculum. Early exposure can also serve as a platform to dispel some of the myths women medical students may have about radiology.

Mentorship – Many of us can trace our career path directly back to one or a couple mentors we had in medical school who were critically important in helping us find our way to the field of radiology. There are myriad opportunities for us to avail ourselves to medical students that we should take full advantage of; from participating in medical student interest groups, to soliciting help with our research and educational endeavors from medical students, to directly volunteering our time as mentors through the school of medicine's mentorship program, just to name a few. Good quality mentorship will lay the foundation for more medical students, and specifically more women medical students, to feel compelled to enter the field.

Role Models – As aforementioned, women medical students tend to go into specialties with higher proportions of female residents. It is completely logical that we tend to follow the path of those to whom we can relate, and look like us. Too often women in the workforce are passed over for promotions and leadership roles not for lack of ability or qualifications, and the field of radiology is no exception. It is our duty as men to not allow our field to continue to be a statistic, and we must make a concerted effort to recruit and make more leadership roles available to qualified women. If we want to inspire and increase the number of female residents in our field, it is critically important that they see more women leaders in the field that will serve as role models.

Our goal as men in radiology should be to challenge the status quo and attract individuals to the profession that are going to add real value, irrespective of gender or any other demographic factor. Everyone benefits when we are all able to pursue subspecialties based on our ability and interest, and no one feels shut out because of who they are. Diversity adds to the richness of our specialty and ultimately has a direct and beneficial impact on the most fundamental pillar of our mission – patient care.

Ibrahim Idakoji, MD, MPH

Clinical Assistant Professor, Department of Radiology
Stanford University School of Medicine

Insights from a Cambodian American Radiologist

I am an Asian American man and am considered a minority in America although not in Medicine. That latter statement is important because while there are many Asians and Asian Americans in healthcare, there is marked heterogeneity within this "umbrella" group. In reality, there is a large education and socioeconomic gap between East and Southeast Asians, which makes our stories different. Although I was born here, my parents escaped the "Cambodian genocide" in the 1970s to start a new life in America. My father, who was once a nurse in Cambodia, obtained a job as a postal worker to make ends meet, while my mother cared for my two siblings and me at home. Growing up, access to resources was not abundant, but my parents were adamant that their children succeeded in life. Because of my background and upbringing, I always felt a personal sense of having something to prove — proving to others that I belonged despite not having as much as them and proving to myself that I could overcome any obstacle in my path, including ignoring those figures who told me that I was not going to make it very far in life. It's this drive that helped me get to where I am today. Fortunately and thankfully, Stan-ford embraces diversity and nurtures the fire within. As such, I feel appreciative to be here and feel the importance of being visible as a Cambodian American Radiologist. Growing up, it would have been great to see someone from my background (or a similar background) succeed in life. This is more important than ever now, so it is wonderful that this newsletter is providing a platform for all of us to celebrate our individual differences and stories.

Michael Iv, MD

Clinical Associate Professor of Radiology
Division of Neuroimaging & Neurointervention
Stanford Medicine | Radiology

Anti-Asian American Violence is Not New

The largest mass lynching in American history took place in the Los Angeles Chinatown in 1871, when 19 people were killed, 4 by gunshot, 15 by hanging, including the only physician serving the community. (https://www.latimes.com/california/story/2021-03-18/reflecting-los-angeles-chinatown-massa-cre-after-atlanta-shootings)

Draftsman and engineer Vincent Chin was murdered in 1982, his skull caved in with a baseball bat by 2 Caucasian automotive workers after he was par-tially blamed for the collapse of Detroit's automotive industry. The murderers were fined $3,000. The Wayne County Circuit Judge Charles Kaufman com-mented "These weren't the kind of men you send to jail." Both murderers were eventually acquitted of Federal civil rights charges. (https://en.wikipe-dia.org/wiki/Murder_of_Vincent_Chin)

Eight people, including six Asian-American women, were murdered in Atlanta on March 16, 2021. Cherokee county sheriff's office Captain Jay Baker ex-plained that the gunman had "a really bad day." The sheriff's Facebook account had a post endorsing the selling of T-shirts with the message "Covid 19 – IMPORTED VIRUS FROM CHYNA."
(https://www.washingtonpost.com/nation/2021/03/17/jay-baker-bad-day/)

White Americans are not referred to as "European," Black Americans are not referred to as "African," but Americans with Asian heritage are referred to as "Asian." This term was adopted as a more neutral alternative to "Oriental," "Yellow," or "Mongoloid." However, it contributes to perpetuate the foreignness of Asian-Americans, and as such, is an institutionalized microaggression. My ancestors were from Asia, but I am not from Asia. When told to go back to where I came from, should I return to Pennsylvania?

Daniel Sze, MD, PhD

Professor
Interventional Radiology
Stanford Medicine | Radiology

In support of

HeForShe

For the third year in a row, the American College of Radiology (ACR) has chosen to support the #HeforShe initiative at the ACR annual leadership meeting.

As a dear friend of mine, Dr. Amy Patel, a breast imager at Beth Israel Deaconess Medical Center and instructor in radiology at Harvard Medical School, explained in her ACR Blog post (https://acrbulletin.org/topics/social-media/1568-heforshe-at-acr-2018):

#HeForShe is a solidarity movement by the United Nations Women which promotes gender equality across all spectrums, including healthcare, education, politics, identity, vi-olence, and work. As a result, this effort has now permeated into medicine with subspecialties such as surgery and radiology taking a stand and promoting gender equality at specialty society meetings and on social media."

During the ACR meeting, members of the ACR Commission on Women and Diversity acted as ambassadors, with signs, stickers and other educational materials available to help inspire awareness, discussion and promotion of the #HeForShe initiative through social media. Attendees were encouraged to take pictures with the #ACR2018 #HeForShe signage and post on social media as a show of solidarity of their underrepresented female radiology colleagues.

As a member of the ACR Commission on Women and Diversity, I was fortunate to be able to work with the talent-ed ACR staff to help organize the initiative this year. In an effort to allow participation from our greater radiology community, the ACR staff developed a tool kit, which included downloadable materials. These materials enabled members of our radiology community who were not physically at the meeting, to demonstrate #HeForShe support from afar via social media channels. I was completely overwhelmed with joy and pride at the tremendously powerful response from the Stanford Radiology community. The outpouring of photographs of staff, residents and fellows demonstrating awareness and solidarity for this movement was truly seen and heard around the county. Stanford Radiology was certainly a strong voice in the #ACR2018 event this year, and I know the women of Stanford Ra-diology and the greater ACR community felt wonderfully supported and celebrated.

K. Elisabeth Hawk, MS, MD, PhD

Clinical Assistant Professor
Stanford Medicine | Radiology

References: The website: http://www.heforshe.org/en
**Emma Watson's incredibly powerful speech at the UN in 2014, such a good listen:
https://www.youtube.com/watch?v=gkjW9PZBRfk
ACR's HeForShe page: https://www.acr.org/Member-Resources/Commissions-Committees/Women-Diversity/HeForShe

Photos from the #HeForShe Initiative at Stanford Radioliology

"Small steps add up to complete big journeys."

Matshona Dhliwayo

"All of our humanity is dependent upon recognizing the humanity in others."

Desmond Tutu

"Whether you think you can, or think you can't — you're right."

Henry Ford

"To survive, men and business corporations must serve."

John H. Patterson

"If I have seen further it is by standing on the shoulders of giants."

– Sir Isaac Newton

The Academic Pipeline

Our next chapter is dedicated to the academic pipeline. Academic pipeline programs and initiatives aim to support and propel students from underrepresented minority backgrounds along their educational journey by providing systematic training in science, technology, engineering and mathematics/medicine (STEM). Pipeline programs provide exposure to clinical observerships, summer internships, research experiences and career development programs. Pipeline programs are important because they help to diversify the academic workforce and introduce the next generation of clinicians and researchers to the STEM field. This chapter provides examples how established researchers and clinicians in the STEM field can introduce young students to the STEM field. This chapter will also provide examples how student interns can provide valuable contributions to an academic Department and the broader community. I would like to thank trainees, faculty and staff who invested significant time and effort to organize student programs, develop student projects and serve as guides and teachers for student interns. I also sincerely thank all sponsors, who are making summer student programs possible and successful.

The student interns who are kindly sharing their experiences will provide the reader with insights about insecurities and fears while stepping into a new world. The student authors explain how some questions or comments can be unintentionally hurtful, how immersion in research experiences can be both challenging and exciting, and how positive feedback and a successful completion of an internship can have a major impact on a students life and career choices. I sincerely thank our student interns for their openness to share their thoughts and giving a voice to others who may feel the same.

To expand this conversation, I would like to cite some of the intriguing questions about academic pursuits and life in general that emerged from our cross-cultural conversations with our trainees, such as: What is the difference between success and mastery? What defines an intellectual? How can we increase our luck? It is a pleasure to see an increasing number of people embarking on a quest for answers to these questions. We are all grateful for the opportunity to learn from each other!

How to be an ally

Few trainees from underrepresented minority backgrounds are entering the field of academic medicine and basic science research, making it difficult to close the race disparity in the STEM field. Early exposure and training in preclinical and clinical biomedical research provides a broad foundation for a wide range of career opportunities, ranging from academic faculty positions, research management positions and careers the industry, among many others. But there is much work to do in order to close current disparities. The points below provide examples, how established STEM members can support the next generation of clinicians and scientists. These points are not meant to be comprehensive and may require adjustments for specific institutions:

- Create outreach programs: Introduce high school students to the importance of post-secondary education and a variety of careers in the STEM field
- Offer paid summer internships and paid student research opportunities
- Offer online interviews and/or travel cost reimbursement for in person interviews.
- Provide incentives and/or appreciation for faculty and staff who invest their time and energy in pipeline programs and mentorship activities
- Do not assign mentors. Provide networking opportunities so that prospective mentors and mentees can interact with each other and choose partnerships based on common interests and shared goals
- Develop a strength-based individual development plan (IDP) with and for every student
- Help the student to build a network of peer contacts, mentors and sponsors. Offer the opportunity to meet a variety of role models from a wide variety of gender and racial/ethnic backgrounds.
- Enable students to strengthen themselves through their own cultural strength: Offer informal gatherings, where participants will receive opportunities to hone storytelling and communication skills, share stories that challenge dominant narratives, create new research ideas, and develop networks of mutual support.
- Provide opportunities for students to regularly discuss and address questions and concerns regarding identity, integration and belonging.
- Offer lectures and discussion groups on resilience skills, such as building confidence, purpose, self-compassion and supportive relationships
- Learn from trainees about their approaches to develop strength and resilience in times of adversities and setbacks
- Provide training for both mentors and mentees in cultural competencies and interactions with people from cultural backgrounds other than their my own
- Provide dedicated lectures on influence and negotiation skills
- Provide input and feedback on the scholar's curriculum vitae (CV) development, and assist with drafting a personal and/or research statement, practicing presentations, or rehearsing interviews
- Provide research opportunities that lead to an opportunity to present at a local, national or international conference
- Provide research opportunities that lead to an authorship on a publication
- Mentors and mentees should be encouraged to stay in contact beyond an internship, in order to enable long-term support and sponsorship

What do I say?

"Go ahead and tell us a little bit about yourself," my mentor said to me. I internally began to panic as four people swiveled in their chairs to face me. What do I say? It was such a simple question… yet I couldn't put the words together to form a coherent sentence.

Maggie Wang (left) and Rachel Huang (right)

I had trouble accepting a large part of who I am. As an Asian American, I had a hard time finding a place where I belonged. When I visit my relatives, my American accent is indication enough that I'm an outsider. However, in the U.S., my appearance, alone, is enough for people to form their own judgments.

But as a member of the lab for the past ten weeks, surrounded by the most understanding, supportive, and diverse individuals, I was able to feel the sense of belonging that I usually only feel when surrounded by my family. It was the environment of acceptance and lightheartedness that gave me the sense of belonging and eventually, acceptance.

"Be yourself and just have fun. Don't try to become someone you're not. Just be you and speak up and be excited," my mentor once said to me to calm my nerves before a presentation—one that ended up being the best presentation I'd ever given.

Originally, I thought I'd leave Stanford having acquired the skills necessary to prepare me for a career in research. But I also learned that by embracing my identity wholeheartedly, not only was I able to push my limits, I was also able to accomplish more.

Different is good. Different is unique and by embracing these differences, we all become more.

Rachel Huang

Home Institution: Johns Hopkins University
Host Stanford Lab: Stoyanova Lab

A Mentor Makes a Difference

Walking into the Canary Center for the first day of my internship, I thought, "Who do I want to define myself as this summer?" I knew because of who I am, others may perceive me as inept or undeserving of my place there. I knew I would deal with microaggression and sexism. I knew I would deal with the inability to relate. I tried not to allow it to foster regret or insecurity. Often throughout my life, I have spent time reminding myself that forces such as affirmative action did not grant me the experiences that I have as a woman of color. I have always worked hard to be successful and to be granted opportunities. I have set myself apart from others. However, being in an environment in which one's integral component of their identity is diminished can foster insecurity. Being a black woman in a field that is predominantly composed of a specific demographic can take a toll on one's self confidence and ability to perform at their best.

As a black woman, often I am aware of my own social surroundings more so than others, because of this, we spend a significant portion of our daily lives socializing and reflecting on our behavior and our interactions with others. We want to feel as though we belong in a place that may not have always been welcoming to people who look like ourselves.

One of the first questions I received was: "So how did you get into the program?" One of the first statements I received was, "wow, how did you get your hair to do that?" I wanted to snarky reply, "I put water in it to make it curly. Just water." However, I didn't. I answered the questions, composed: "I was selected" because I was. It became prevalent to me that the rumors regarding diversity in STEM were true. The Canary Center itself is not a force that breeds micro-aggressive rhetoric, still, I couldn't escape, the questions and the exclusion. I began work in the Gevaert Bioinformatics lab at the Stanford School of Medicine. Immediately, I delved into the world of science. Research and intellectual progression are blind to color and are blind to gender. One can engage with information and learning, because knowledge does not discriminate.

I was assigned to my very own post-doc, Dr. Hong Zheng, who would guide me and teach me mechanisms of research as well as tools for learning coding languages. Her patience and her willingness to put in time into my progression was admirable. Her successes and her knowledge were impressive to me. I consider her to be one of the first female mentors I've had in the STEM field. I believe after ten weeks were completed that I was able to be challenged and to be successful without fear of marginalization. I felt as though I was able to feel like I belong, with my mentor challenging me and inspiring me, I was able to gain valuable experience, free of ridicule due to the color of my skin.

Irmina Benson

Home Institution: Wesleyan University
Host Stanford Lab: Geveart Lab

Realize the Humanizing Side of Research

After finishing my first year of medical school at the University of Alabama in Birmingham, I had a basic understanding of the field of radiology. In a search for furthering my understanding and appreciation for the field, I joined Dr. Daldrup-Link's lab research on the effectiveness of PET/MR in the local pediatric population. I gained a breadth of knowledge involving the intricacies of imaging during my time at Stanford, and was fortuitously given the chance to observe many of the scans that would be used in the research that I was involved in. There was an incredible diversity in the patient backgrounds in the Bay Area, and I had the exciting opportunity to speak with and for a Hispanic family during their daughter's imaging scan. As a son of two Hispanic parents involved in improving healthcare access to Latino populations in Birmingham, I was excited to do so in a way that is unique to my goals as a physician-scientist. While at Stanford, I saw the humanizing side of research, where my work in the lab directly affected the lives of the patients in front of me and future patients with similar afflictions.

Jordi Garcia-Diaz

Home Institution: University of Alabama School of Medicine
Host Stanford Lab: Daldrup-Link Lab

Numbers that fuel passion and dedication

Diversity means more than just the differences in our cultures and our ethnicities. When I think of diversity I also think of the diversity of our experiences that make us who we are. These experiences are not evident by the color of our skin or our last names. It requires digging deeper. If you looked at a profile of me you'd see my name, age, date of birth, MCAT scores, Step 1 score, and where I went to college and medical school; however, these are not the experiences or numbers that make me who I am or that inspired me to dream. The numbers that fueled my passion and dedication are the numbers that I could not bubble into an application. I am the fifth child of a family of seven children and two parents. My two parents immigrated into this country 36 years ago and in that time have been unmotivated to find stable jobs, learn English, or integrate successfully into the United States. My father resorted to illegal means of making money and landed in jail two times before I was six. My parents relied on state aid to raise my siblings and I, and words like welfare and food stamps were part of my lexicon from a very young age. My mother had my youngest and only brother when she was 39; he was born with Klinefelter's Syndrome, meaning he has one extra X chromosome that delays his mental and physical development. I realized at a young age that life had given me some challenges: the first challenge was my ability to dream beyond my circumstances. I had to fight feelings that I was unworthy of the opportunities before me and battle against the statistics of those who try to escape poverty but are unsuccessful.

I am proud to say I will be applying into residency for diagnostic radiology this year and will fulfill my lifelong dream of becoming a physician. This is a great achievement and I am very proud of myself; however, this achievement along with others that may come in the future, are meaningless if it means that I attend residency, obtain a prestigious job, and then forget. Forget where I came from, how I got here, and forget to give back. I have been given so many challenges in my life and struggles at a young age, but most importantly, I have been given the strength and ability to see myself beyond my circumstances. Because I have been given this, I know that much is expected of me as a member of my community, and I am expected to give back to those who also have challenging numbers.

Imilice Castro Paz

Hometown: Riverside, CA
Home Institution: Stanford University School of Medicine
Host Stanford Lab: Dr. Terry Desser & Dr. Daniel Rubin

Knowing your Purpose Defeats Imposter Syndrome

When I came to Stanford, I can admit that I felt a huge case of imposter syndrome. I was chosen to be a part of a small cohort for the summer, but I still felt I was not as brilliant as the other students around me. Luckily, I was placed in the Daldrup-Link lab this summer, and I adjusted well to the atmosphere and the research. While working on a diversity project, I was able to interview faculty members in the radiology department, and I learned that even well accomplished faculty can still feel imposter syndrome at times. While I was curious to why this may be, I realized that I was in a similar situation. From that point on, I was able to finish my internship knowing that I was here for a reason and that I was just as good as everyone else. I feel this is something all faculty members should own about themselves. We are all placed where we are now for a greater purpose. You may not see it immediately, but with time, you will know what that purpose is.

Kensley Villavasso

Home Institution: Xavier University of Louisiana
Host Stanford Lab: Daldrup-Link Lab

Not All Life Saviors Need to go to Medical School

I, Lilly Zhou, am a prospective computer science major in the artificial intelligence track. This summer I was interning at the Radiological Science Lab (RSL) at the Lucas Research Center at Stanford as a software researcher. I worked to develop a mixed-reality system that utilizes object tracking cameras and Microsoft HoloLens to project a 3D "hologram" of images from a preoperative breast tumor MRI onto a patient. My project aimed to improve surgeons' ability to determine the location of tumors during surgical operation. During the internship, I was amazed to see the broad spectrum of ways computer science can be integrated into many different fields including surgery and medical imaging. If one is interested in both healthcare and computer science, like myself, the good news is they do not have to choose between the two because it is by all means possible to do both. One important thing I learned during my time there is that not all life savers necessarily need to go to medical schools. One can also make their own unique impact on other people's lives by building things and solving problems outside the hospital buildings.

Lilly Zhou

Home Institution: Stanford University (Rising Sophomore)
Host Stanford Lab: McNab Lab

Investing Extra Work to try a New Approach will Make you Proud

My name is Akshay Jaggi, and I am a rising senior at Stanford. I major in Biology (honors track in computational biology) with a double minor in linguistics and computer science. I'm also on a competitive Indian dance team on campus called Basmati Raas (pictured below). I am interested in applying natural language processing and machine learning to improve medical diagnostics. This summer, I worked with Dr. Sandy Napel on further developing digital phantoms for image feature quantification standardization. I completely reworked an approach to the problem that I developed last year. This summer, I certainly learned that you're never too deep down a path to reevaluate your decisions and develop a new approach. That new approach may require much extra work, but, if the outcome is a project you're more proud of, then the work is worth it. A mindset that will hopefully be useful during graduate work I pursue in the future.

Akshay Jaggi

Home Institution: Stanford University
Host Stanford Lab: Sandy Napel Lab

A Setback is not a Failure if it is Temporary: Trudge through Difficulties to Reach Success

I have always been passionate about academic medicine and have hoped to find opportunities to practically apply this desire to help advance the field. Clinical research allows this very ambition to be possible, but it always seemed so broad and intimidating to me, especially as an undergraduate student. However, after my summer in the Daldrup-Link lab, it appears that research is not only accessible, but also manageable with the right opportunities and mentors. Each researcher in the lab, especially my mentors Dr. Anne Muehe and Dr. Daldrup-Link, has taught me what it means to be a researcher and a physician. I've had opportunities to marvel at new devices for nanoparticle labeling, observe a porcine knee surgery, learn the basics of cell culturing, create artistic medical figures by hand, write a review paper, and to be bold. Each member of the lab has continually encouraged me to be confident in my work, even when I make mistakes. When I found myself constantly editing and re-drawing my figures on photoshop or stumbling to present at a lab meeting despite days of preparation, I felt utterly defeated. Despite these frustrations, I was always told the same thing at the end of every meeting: "Good job!" Those words have motivated me each day to wake up at 7:00am and to have a researcher's will to trudge through difficulties to reach success. I have also gained an invaluable understanding of an academic physician's role in promoting scientific discoveries. These new perspectives have made each day at the lab more fun than the last, and I hope to continue developing my passion and skills to become a physician scientist!

Jessica Tseng

Home Institution: University of California, San Diego
Host Stanford Lab: Daldrup-Link Lab

Research Requires many Re-Searches

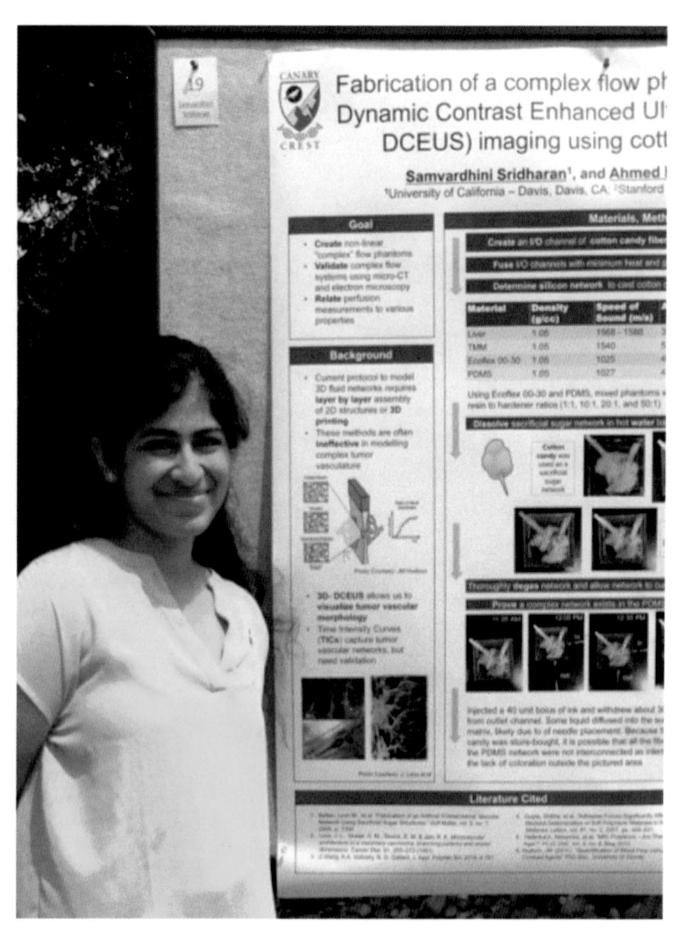

I thought Murphy's Law's association to research was just an exaggeration: The gravity of science, ultimately, is its tendency to bring complex questions down to Earth. Only when responsible for a summer project of my own did I realize that my previous naivety knew no bounds. Research, after all, is full of problems.

This summer, I've been able to demystify what was once a fancy word. "Research" isn't a puzzle that I assemble piece by piece. Rather, it is put together and taken apart, talked about, modified, and then reassembled some more. A single misplaced piece has the potential to not only tumble an enigma down, but to also metamorphose it to become a better version of itself.

Learning what research is about meant modifying what is the best way to "do" science. Inductivism is attractive in theory, but I've surmised it hardly works. This is perhaps the greatest learning experience that I will carry with me: research hardly works. This summer has been heuristic, and Mr. Murphy and I have become good friends. As I pursue research, I'll allow time for his law to do its job – the most fruitful conclusions are left in its wake.

Samvardhini Sridharan

Home Institution: University of California, Davis
Host Stanford Lab: Willmann Lab

'What-if' Queries can lead to New Insights

Cameron Moseley working on his review article

As a Persian-American high-schooler with a unique opportunity to work with Dr. Spielman in Radiology, I have acquired a keen sense of what 'clinical research' means now. This was never a topic in any of my studies at Leland, and would have most likely never been much in my vision of my future. I understand now the challenges of translational research, how 'what-if' queries can lead to new insights and how something as mysterious as 'amplified MRI' could affect everyday life. Over the last two summers, I got a special glimpse into how Radiology research works, how working at the Lucas Center led me to many interesting people and ideas, and how this work could become my future. As a result of my experiences and interactions with a lot of cool people at Lucas, I have decided to focus my college to a 'pre-med' line of studies. I would like to mention those who were so kind to help me; Drs. Tom Brosnan (thank you Tom for your mindfulness chats), Ashok Theruvath, Anne Muehe, Dan Spielman, Samantha Reyes, and Mehdi Razavi. Thank you to everyone at Lucas as well; it is a rich and wonderful environment.

Cameron Moseley

Previous: Leland High School, San Jose
September 2018 Institution: University of California, Davis
Host Stanford Lab: Spielman Lab

Laughs, learning experiences, spontaneous adventures, and amazing memories

Never have I felt so at home within a scientific community. As a first-generation scholar from Central California, I was fortunate to be funded by the Amgen Foundation to conduct summer research under the Department of Radiology at Stanford. A microbiologist and bioinformatician by trade, my project challenged me to explore biophysics as I modeled the uncaging mechanisms of nanoparticle drug carriers. As a member of the Amgen Scholars Program and Stanford Summer Research Program (SSRP), I experienced the life of a graduate student for two months as I was immersed in science, evening seminars, workshops, and GRE prep. My passion for translational research was amplified by my short-time in the Airan lab, solidifying my decision to pursue a graduate degree in biophysics. Thank you to Dr. Airan, my mentor Sunmee Park, the Department of Radiology, and my SSRP family for making my summer 2018 filled with many laughs, learning experiences, spontaneous adventures, and amazing memories.

Brenda Yu

Home Institution: University of California, Merced
Host Stanford Lab: Raag Airan Lab

Science: Sincere Inquiry, Rigorous Scrutiny and Learning from Diverse Situations

I had the amazing opportunity this summer to work with Stanford Radiology building artificial intelligence tools for the analysis of deep brain structures such as the thalamus within MRI images. Coming from a background in theory, I was extremely excited to apply deep learning methods to a real task in the medical community that has the potential to generate new insights into devastating neurodegenerative disorders. Through working with professors in the department as well as spectacularly talented graduate students and fellows, I was able to make progress towards my goal of being able to automatically detect regions of the brain frequently affected by diseases such as Alzheimer's. However, working with Stanford Radiology taught me something more fundamental to my life and career; I learned what it truly means to be a scientific investigator. Science as a whole greatly outweighs the sum of its success—it is about living a lifestyle that facilitates sincere inquiry alongside rigorous scrutiny and documentation. It is about waking up early to watch brilliant presentations, staying late into the night in pursuit of knowledge, treating failures with the same excitement as triumphs, and never failing to learn from diverse situations. Thank you RSL!

Matthew Dardet

Home Institution: Stanford University, B.S. Candidate in Mathematics
Host Stanford Lab: Rutt Lab

*" We make a living by what we get,
we make a life by what we give."*

Winston Churchill

"The delicate balance of mentoring someone is not creating them in your own image, but giving them the opportunity to create themselves."

Steven Spielberg

"A mentor is someone who allows you to see the hope inside yourself."

Oprah Winfrey

"As iron sharpens iron, so one person sharpens another."

Proverbs 27:17

Notes

FIRST GENERATION

"It doesn't matter where you came from, all that matters is where you're going."

– Brian Tracy

First Generation

This chapter is dedicated to first generation faculty, trainees and staff in the STEM field. This topic is particularly close to my heart as I, too, am a first generation medical doctor in my family and a first generation American. It has been an exciting journey and I am beyond grateful for the many opportunities that I have been given along the way.

I noticed that the term "first gen" is sometimes seen as a deficit, rather than strength. For first gen graduates, the entire cultural context of a University might be foreign, and many might face financial constraints. At the same time, first gen's might have extraordinary abilities with regards to their determination, resilience and creativity. First gen's often have been trained their entire lives in finding solutions to apparently insurmountable obstacles. First gen's are risk takers, "out of the box" thinkers and problem solvers. As a community, we have to enable them to contribute their unique skills and abilities.

For some first gen's, entering an elite University means bring-ing honor and pride to their families. It might reflect the strength, hard work and tenacity of the entire family. Many first generation graduates have experienced struggle and suffering. Arriving at a University like Stanford can feel like reaching a clearing at the end of a long, hard journey. To these colleagues, I would like to say: We are proud of you! Keep moving forward! Being at an elite University is a condition, not an achievement. Being at an elite University means access to resources, abundant opportunities, making the next great discovery and making a difference in other people's lives. We are humbled, grateful, and blessed!

I remember that stepping into a new world of apparently abundant resources and opportunities was coupled with high expectations for the people I would encounter. In my younger self 's mind, these people were smart, strong, integer, courageous, compassionate and faithful. Unfortunately, during my career path, I had to make some different experiences. But I still believe that true leaders are like that. I hope that readers of this chapter will strive to live up to this expectation – that we will enable everyone to make meaningful contributions, that we will support each other and become happy when we see each other grow. We need an orchestra to play our symphony.

How to be an ally

Many of us are struggling with what we can do as individuals in a political, biased environment, where racism and xenophobia remain perversive. Here, we would like to share a few actionable items that worked for our community. These points provide examples, are not meant to be comprehensive and may require adjustments for different communities and contexts:

- First gen team members come from a wide variety of backgrounds. Do not impose any presumptions or stereotypes.
- Introduce an administrative point person or office that first-generation students can contact easily and informally for any questions or concerns
- Create an on-boarding document with detailed information about your institution, department and community.
- Create a self-assessment questionnaire for new team members, which will help to identify their professional goals as well as individual strengths and challenges.
- First gen's often possess highly desirable traits for the STEM field, such as grit, self-determination, problem-solving skills and important insights into health disparities. Provide first gen's with opportunities to share their insights and ideas with regards to practices and workflows at your institution and community outreach
- Develop a first gen mentorship program. Connect first-gen scholars at your institution. Offer networking opportunities with peer mentors and senior sponsors.
- Distribute "First Gen" or "I'm First!" pins to recognize and honor the accomplishments of first-gen college graduates
- Consider the existing social network of first gen scholars an intrinsic asset that can be a source of information and possible bridge for outreach efforts
- Create a list of language skills of your team members.
- Make career development opportunities available to first-gen scholars, such as books and seminars on career paths; career counselors; sponsors and related resources of professional societies
- Organize an imposter syndrome workshop
- Offer leadership training through educational lectures/seminars or a leadership coach
- Provide lectures for teachers on inclusive teaching, considering diverse backgrounds, how to create an equitable environment, and how hidden curriculum influence students' learning and participation
- Check your meeting minutes: Whose voices have you captured? Did your meeting involve a discussion or a monologue?
- Introduce formal processes to address microaggressions and tone policing
- Provide access to wellness programs and mental health counseling programs
- Generate a list of resources, scholarships, stipend and grant opportunities for first gen scholars at different career stages, to address any financial constraints
- Actively check in with scholars in regular intervals and provide career counseling services to address any challenges as they arise in order to minimize drop-outs
- Organize information sessions, social events and activities that invite and welcome family members of first-gen students to the institution.
- Administer a yearly climate survey in order to collect information about potential areas where bias or prejudice might impair the work or career progress of first gen scholars

Barriers to the Success of First Gen Students

I don't think I've experienced any cultural barriers impeding my long-term success, however, I have faced academic barriers that are pretty common among FLI (first-gen and/or low-income) students.

The difficulty of introductory courses in STEM fields, such as General Chemistry and Organic Chemistry I, despite there being additional courses to take alongside chemistry courses to provide extra practice sessions, these STEM courses are quite simply weed out courses. Students who have not yet been able to learn how to study and who have not faced this level of rigor in classes before are expected to transition almost immediately to this rigor. Even though I went to office hours, took the extra courses for additional practice, talked to the professors and the TAs, I always felt like I was playing catch-up. From my personal experiences, most of the people I know who stopped being pre-med in their first year were FLI brown and black students.

After joining our lab and having Dr. Kiru as a mentor, I finally came to realize how important it is to have someone that looks like you in academic and research settings. In my 6 quarters at Stanford, I have had 1 black professor. Although this doesn't pose a barrier necessarily, I think it would be beneficial to the university to consider employing more black and brown professors. As a FLI student, it can sometimes feel that you're the only person struggling. Having professors and mentors that look like you can give you affirmation that your struggles are not silent because, more often than not, they faced similar barriers.

Although there are ways Stanford can progress, overall, I think that programs such as Leland Scholars Program to help FLI students transition to Stanford before freshmen year starts have been really amazing for black and brown undergraduates. I also think the cultural centers have served as needed safe spaces for students of color. Thank you!

Famyrah Lafortune

Undergraduate Student
Stanford University

"I believe race is too heavy a burden to carry into the 21st century. It's time to lay it down. We all came here in a different ships, but now we are all in the same boat."

– John Lewis

"Talent and effort, combined with our various backgrounds and life experiences, has always been the lifeblood of our singular American genius."

– Michelle Obama

Insights from first Generation Graduates at Stanford Radiology

"You can't use up creativity - The more you use, the more you have."

– Maya Angelou

Insights from first Generation Graduates at Stanford Radiology

I am the first PhD in my family. My parents get a kick out of calling me "doctor" and always say it with special emphasis. I'm sure they still see me as the goofy little kid singing and dancing around the house. My parents were incredibly hands-off when it came to school and career. The only advice I got was "do your best" and "I'm sure you'll find something you love to do." Picking a major, applying to grad school, completing qualifying exams, and preparing for my doctoral defense were 100% on my own. My family loved and supported me but had no idea how to guide me through these milestones. I'm still not sure they understand what a postdoc is or why I'm doing one. And that's okay. They are tremendously proud to say that I have a PhD and work at Stanford and I feel honored to set the precedent for my loved ones.

– Anonymous

I was born in Caborca, Mexico and was fortunate enough to have parents who made endless sacrifices to ensure my three siblings and I could have the best opportunities to succeed.

Soon after I was born, my dad moved to the US and sought work as a day laborer doing construction in Phoenix. Back in Caborca, my mom and my siblings and I lived with relatives and struggled to make ends meet. When I was 11, we crossed the border and joined my dad in Phoenix. My parents both continued to work long hours in physically demanding jobs – my mom as a housekeeper and my dad as a landscaper. Fortunately, our socioeconomic status slowly improved; my dad became a citizen, the rest of us achieved permanent resident status, and my dad started his own landscaping company. My parents set me up to be the first person to attend college and subsequently medical school. I am indebted to them and hope that I am now in a position to serve as an advocate for greater diversity in our program, recruit minority applicants, and mentor diverse students who are interested in radiology.

– Mario Moreno, MD

I can complete the phrase "first generation" so many ways – Greek-American, college graduate, physician, interventional radiologist. As typical for someone coming from an immigrant family, much of my childhood and young adulthood was spent trying to find my place between my very Greek parents and "those Americans" (as my parents called them). I was made fun of for bringing spanakopita for lunch as a kid and for not knowing common English words that my parents never used at home. Later on, I felt out of place because I didn't have a parent as an alumnus of my undergraduate or medical school. However, I feel so fortunate to have my parents motivating me and supporting me every step of the way. As I grew older, I molded my own identity. Now, while I still encounter awkward moments that remind me of the cultures I straddle, I am honored to help other first generation physicians along their path. I also would be thrilled to eat spanakopita for lunch every day.

– Amanda Rigas, MD

Insights from first Generation Graduates at Stanford Radiology

Growing up in the Southwest and East Coast, looking different from pretty much everyone else was strange to say the least. I was a brown kid, in a very un-brown neighborhood.

While most of my childhood was benign, there were unfortunately numerous periods punctuated by actions intended to hurt me. Actions that left me feeling deeply alone, angry, and outright terrified. Sometimes I found myself at the bottom of a dark pit, completely paralyzed by self-hatred and utter confusion. After all, I was a quiet kid who went about my business. Why did they treat me this way? Why did they feel compelled to gang up on me? And why did I feel so helpless? I would just try stay quiet and still. Walking away would only provoke them to physically attack me. I was trapped until they tired. I hated being brown.

In elementary and junior high school, for example, it was not unusual for some of the school kids to mock my skin color by performing a poor rendition of a Native American pow wow dance around me, or call me cruel nicknames, likening my color to that of feces. It pains me to say it also wasn't uncommon to be dragged into yet another fight to defend my little sister's honor when several boys would make inappropriate gestures at her. Or when I was routinely picked last to be on a sports team because there is no way a brown Asian kid can be good at sports. Or even being told by the High School Principal that I did not deserve to be Valedictorian because there were smarter (white) students than me in the school, even though being the 'top student' in your class was defined by your GPA. "Really??!?" I would say to myself, shaking my head and feeling sick to my stomach. There were also some rather NSFW expletives I said in my head when that happened—I'm still in disbelief when I reflect upon these experiences. They make me think of all the people of color across the globe, and those belonging to different sexual and gender minorities, who share in this disturbing reality on a daily basis. Many of which who have it a LOT worse than me. These thoughts make my soul cry and despair, and at the same time give me strength when I realize the amount of courage they must muster, just to survive.

For me, surviving was made possible only through the loving support of my parents. I was also fueled by a personal desire to prove those haters wrong. Damn it if I'm gonna let them win, I often said to myself (and still do to this day). I need to prove everyone wrong, and that I'm someone who can do something meaningful. Turns out I did a few O.K. things over the years...First place in the City Science Fair, recipient of some Merit Awards, entrance into a few good schools and fellowship programs, Teacher of the Year awards, and now a faculty member, part of the Stanford family. I do, however, walk into life every day with an undeniable chip on my shoulder.

Is that a good thing? Do I even dare say that perhaps there is silver lining to being treated so poorly in the past? I'm not sure I would go that far, because I wouldn't want anyone to experience what I did. Fortunately for me, I had a tremendous network of support around me, including my parents, siblings, friends and teachers who gave me the foundation and confidence to overcome those experiences. But there is certainly a fire that burns deep inside me to this day, to do something positive and help empower those around me, especially those less fortunate. That fire came from a not-so-good place. However, I am thankful that I have the opportunity to use what I've been through for something good. At least, that is my hope every day.

– Anonymous and Edited by Anonymous's Close Friend

What First Gen Graduates Hear

When will you earn money?

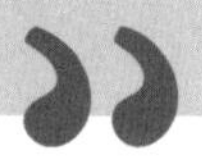

“

I have exciting news!!

— A baby??

No, I passed my exam!

— Oh.

“

Dad, how can I solve this problem?

— You are so smart.
You will figure it out.

Mom looking at my new badge:

You look fantastic!

Dad, my friends want to attend that overseas program.
It costs $3,000.

—*I guess that means you stay here?*

What do you want to be in 10 years?

— *Alive!*

The Pursuit of Proactive (Mental) Health

A few of my colleagues heard me speak about mental health, stress, and affective technology and heard me say: "the pursuit of anything is stressful." I said this in light of people asking me about happiness, and how I believe the constitution of the US is a stressful document, as it promotes the "pursuit of happiness". However, people do not always understand that something stressful is not necessarily a bad thing. Stress needs not to be eradicated, but it definitely needs to be managed - too little or too much stress can lead to serious physical and mental health implications. In this short reflection about my "pursuit" to collaborate with my colleagues in the creation of proactive (mental) health in science, I will discuss three key topics that evolved during my career as a "stress management" technology designer and researcher: I will discuss (1) the need to use design practices to develop products that could help out large populations to stay healthy, (2) the need to influence the public opinion around the need to focus on mental health, and finally, (3) the need to create new markets for proactive mental health technology. This is, by no means, an exhaustive list of relevant topics around mental health, but these are the topics I believe I could impact.

Proactive mental health design for everyone

My research lies in the chasm dividing mental health research from technological advances. I want public and mental health researchers to use technology that is not 30 or 40 years old. Motivated by my experience with caring for a family member who was diagnosed and treated late of a debilitating mental disorder, I decided to pursue a career in science in order to give other families better chances of early detection and management. My journey started by reading the Larousse Encyclopedia back in Ecuador, long before Google was invented. After reading about anxiety and depression, I realized that my family member was not just behaving as a difficult teenager. Many years later, and after many "kilos" of medicine consumed by my family member, I gave up a successful engineering and business career on transforming the future of mental health technology. Mental health affects roughly 25% of the population, and it generates about $1 Trillion in losses worldwide [2]. As stated in my recent TEDx talk[i], I estimate that the efficiency of mental health therapy, i.e., its ability to help this one-quarter of the population, is less than 0.5% [1,3,4], and it is mostly a result of low adoption and high attrition. Ultimately, I want to invent scientifically-backed, proactive, and engaging mental health technology to close the gap of mental health care for the masses.

Influencing opinion about mental health

My faculty position within the Precision Health and Integrated Diagnostics Center (PHIND)[ii], one of the most emblematic projects of the Stanford School of Medicine (SoM), has given me the ability to influence the discussion on how to move from a disease-centric medicine model to a health-centric wellbeing model. Through enriching, and at times argumentative, conversations with PHIND's leader, and my boss, Dr. Sam Gambhir, I have developed a finer sensitivity about the need to know, based on hard data, what it means to be healthy, and how to keep people that way. In recent months, my work has been portrayed in the SoM's monthly magazine, describing the synergistic use of design-thinking and the scientific method to create engaging technology for healthy people, certainly a non-zero-sum game opportunity. I have started conversations with key influencers such as Dr. Robert Sapolsky and Dr. Victor Carrion to help me craft a message for adopting a culture of stress "hygiene" - a topic not explored in the health industry. Recently, I have been generating evidence to motivate federal agencies

to fund more work in proactive IoT for health. Despite a very limited focus on proactive health research, Program Officers for the Smart and Connected Health (NSF/NIH) and Cyber-Human Systems (NSF) grants, and the DoD have expressed some interest in my ideas and papers.
Given the stigma associated with mental health, I find it advantageous, and at times an obligation, to leverage my strong communication skills to change public opinion. My scientific status allows me to deliver a compelling message through multiple outlets. Recently I have spoken at the Brain Mind summit, an event bringing scientists and venture capitalists (VCs) together, organized by Stanford and MIT. I have also proposed my ideas around IoT for proactive health led invitations in forums such as the Transformational Technology (TRANSTECH) industry conference in Palo Alto and a TEDx talk in Boston,

Developing mental health technology markets

A fundamental aspect of new science is the effect, or potential effect, it can have on new market creation. Here is where I wear my prior hat as a strategic manager and blend it with my recent scientific experience to foster the creation of new markets. Given the rather revolutionary vision of proactive health, I like engaging with both large and small companies interested on wellbeing. For example, Habits, a startup company focused on behavior change in the workplace, has invited me to be a board member to create an app for the Spanish-speaking workforce in Latam and the US. On the other hand, large impact entities such as AAA are discussing with us how to offer solutions for wellbeing based my low-cost stress sensing and interventions technology for commuters. They are interested in "productizing" some of our inventions to offer it to their close to 60 million members as part of their community engagement plans. In parallel, I mingle with potential sponsors. I have engaged with VCs, such as Varvara Russkova from GVA Capital, who wants to explore new investment portfolios for proactive health inspired in my work on subtle interventions for the office. My hope is that at some point I can align the different forces to sparkplug new markets and new scientific endeavors.
So... despite being a junior scientist, my past experience as an entrepreneur and businessman mixed with my communication skills and an untamable passion and creativity for design and science fuels my never-ending "pursuit" for the creation of proactive mental health technology. Perhaps my tenacious attitude will draw enough attention from the public, the government, and industry to support this journey to solve the mental health gap for future generations.

Pablo Enrique Paredes Castro, PhD

Instructor, Radiology and Psychiatry & Behavioral Sciences
Stanford Medicine

References

1. American Psychological Association. 2010. Stress in America Findings. America: 64. Retrieved from http://www.apa.org/news/press/releases/stress/national-report.pdf
2. Dan Chisholm, Kim Sweeny, Peter Sheehan, Bruce Rasmussen, Filip Smit, Pim Cuijpers, and Shekhar Saxena. 2016. Scaling-up treatment of depression and anxiety: A global return on investment analysis. The Lancet Psychiatry 3, 5: 415–424. https://doi.org/10.1016/S2215-0366(16)30024-4
3. David C Mohr, Stephen M Schueller, William T Riley, C Hendricks Brown, Pim Cuijpers, Naihua Duan, Mary J Kwasny, Colleen Stiles-Shields, and Ken Cheung. 2015. Trials of Intervention Principles: Evaluation Methods for Evolving Behavioral Intervention Technologies. Journal of Medical Internet Research 17, 7: e166. https://doi.org/10.2196/jmir.4391
4. Ricardo F. Muñoz and Tamar Mendelson. 2005. Toward evidence-based interventions for diverse populations: The San Francisco General Hospital prevention and treatment manuals. Journal of consulting and clinical psychology 73, 5: 790–799. https://doi.org/10.1037/0022-006X.73.5.790

[i] https://tedxbeaconstreet.com/speakers/pablo-paredes/
[ii] https://med.stanford.edu/phind.html

Examples of Famous First-Gen Graduates

Walt Disney	Albert Einstein	Ruth Simmons
Larry King	Steve Jobs	Clarence Thomas
Howard Schultz	Oprah Winfrey	Tom Bradley
Jimmy Carter	Michelle Obama	Henry Louis Gates Jr.
Gerald Ford	Hillary Clinton	Ursula Burns
Richard Nixon	Bill Clinton,	Viola Davis
Sonia Sotomayor	Margaret Thatcher	Ron Dellums
Thurgood Marshall	Ruth Bather Ginsburg	Brian Greene
Ken Lagone	Elizabeth Warren	Samuel Jackson
John Lewis	Colin Powell	Viet Nguyen
Laurie Richer	Marc Tessier-Lavigne	Brian Geene

"Human progress is neither automatic nor inevitable... Every step toward the goal of justice requires sacrifice, suffering, and struggle; the tireless exertions and passionate concern of dedicated individuals."

Martin Luther King, Jr.

Notes

AGEISM

"Life isn't about your age. Life is about living. So, when your birthday comes, be thankful for the year that has just past and anticipate with a happy heart what the coming year will bring."

– Catherine Pulsifer

Ageism

This chapter should initiate a discussion about age. This includes insights and reflections from physicians, trainees and staff who might feel young, old or anything in between. The interested reader will learn about aging in different cultures, journeys through Academic Medicine, a note from a former Chair of the Nobel Assembly to his 30 year old self, and why experience in medicine matters.

On behalf of our STEM community, I want to take this opportunity to express our collective appreciation and respect for our senior colleagues and staff, for their contributions to our healthcare teams and science think tanks. Ageism in a workplace describes prejudice or discrimination on the grounds of a person's age. Examples include stereotyping older team members as being resistant to change or slow, passing over older employees for career opportunities or focusing workplace policies and politics on the needs and wants of younger workers, without considering the needs of older team members. Interestingly, it is often not until a senior employee leaves that a team realizes that it will take two or three new hires to meet the requirements of that position. The field of medicine specifically is highly dependent on professional experience. To provide the best possible care for our patients, it is of utmost importance for our community to enable and retain the decades of procedural and organizational expertise and wisdom of our senior team members.

Time is fascinating and precious. Every one of us was once the youngest person on earth. And when the oldest person on earth was born, there was an entirely different set of people on it. Our internal and external clocks are running and do not pause for anyone. Sometimes, a moment can feel like an eternity. Sometimes, years seem to pass by in an instant. Albert Einstein concluded: *"Not everything that counts can be counted"*.

How to be an ally

Ageism is unique compared to other forms of prejudice and discrimination, because most of us will find ourselves sooner or later in the group of older people. Unlike race and sex, age is experienced by almost everyone. However, many people associate aging with negative stereotypes and this can result in inadvertent biases or microaggressions. The examples below demonstrate how people of all ages can support their senior co-workers. As with all topics discussed in this book, the points below are not meant to be comprehensive and may require adjustments for specific situations:

- Educate yourself about bias and discrimination against older workers.
- Do a thorough self-examination: Do you have a positive outlook on aging and what are your associations with the aging process? Do you project these on other people?
- Be a giver: When interacting with senior colleagues, some people show a taker mentality, expecting personal gain from every interaction. Be the opposite: Plan to add value to every interaction.
- Create intergenerational programs and seminars, which bring people of different ages together for mutually beneficial discussions and professional activities
- Break the stereotype that sponsorship is a one way street: Be a role model in publicly supporting a colleague or team member who is older than you.
- Do not paternalize older people by making decisions on their behalf or acting for them because you believe they are too weak or incompetent.
- Assume all people are mentally and physically able until informed otherwise
- Do not infantalize older people by speaking with them in a child-like tone, slower than usual or using simplified language.
- Do not dismiss an older patient's concerns or symptom simply because he/she is old
- Make sure to include junior, mid-career and senior team members in team meetings and decision-making processes
- Include your senior team members in social media posts and news coverages
- If you show photos or sketches of people in your presentations, show people from a wide variety of demographic backgrounds
- Speak up if you see someone discrediting an older colleague's work or comment
- Publicly recognize and appreciate what you learn from your older coworkers.
- Do not deny time off for family commitments because a worker is older and does not have young children.
- Make sure to document any implicit biases and statements about someone's age or incidents of younger colleagues being treated differently than an older co-worker. Document dates, times and witnesses to conversations.
- Introduce a peer learning conference for cultural competency and interpersonal communications where incidents of bias and microaggression can be discussed
- Offer training on new skills and new equipment to ALL employees
- Eliminate stacked rankings: Instead of accurately measuring an employee's performance against an objective standard, ranking employees according to ill-defined, subjective criteria promotes ageism, derails teamwork and pits employees against one another
- Be an ally at home: Be a positive role model for your family members. Model respect and positive associations with aging for younger members of your family

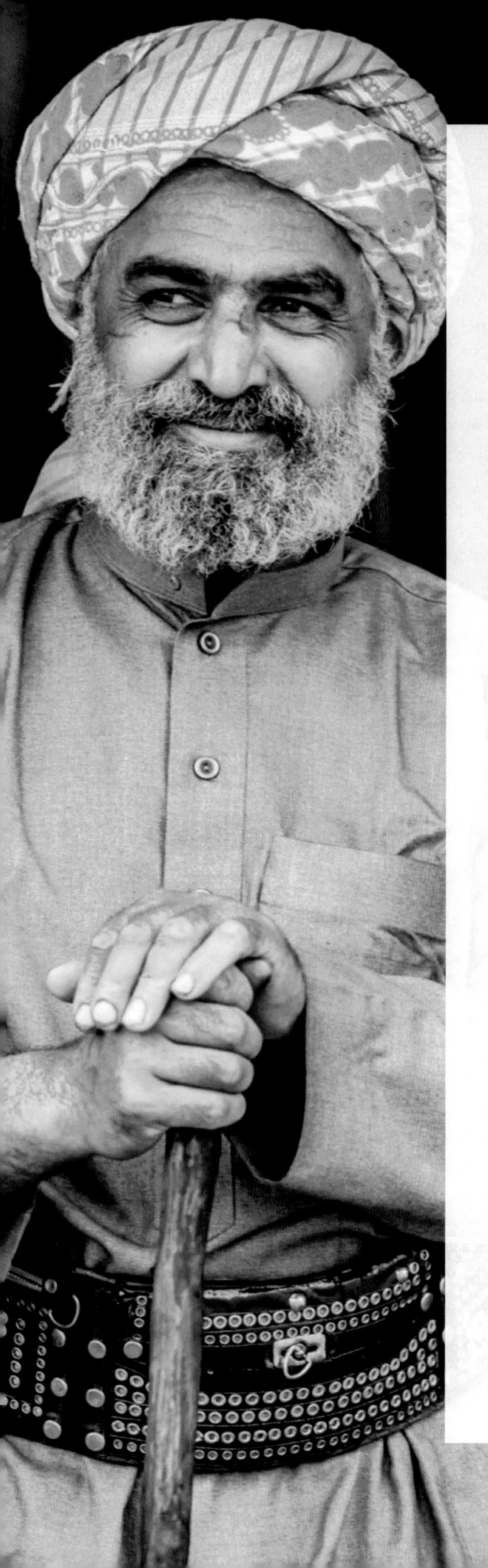

Ageism: Stereotyping and Prejudice against Older Persons

A book Edited by Todd D. Nelson. MIT Press, 2002. ISBN: 0-262-14077-2 Book review in the N Engl J Med 2003; 348:89-90; DOI: 10.1056/NEJM200301023480125; Commentary in Medscape: https://www.medscape.com/viewarti-cle/894078

Unlike racism, sexism, and homophobia, ageism represents a prejudice against a group that all members of the "in" group will inevitably join if they live long enough. Ageism is a phenomenon that allows us to maintain negative stereotypes about our future selves. What vision of old age could have led to society's view of the traditional nursing home as a fitting and pleasant solution to the health problems of old people? Surely, it was not a vision of old age as a vibrant period or a vision of old people as important parts of their families and communities.

Ageism involves negative stereotypes about seniors, which are rooted in the fear of death. Older people themselves may share ageist views of their contemporaries and of those who are slightly older than they. Such implicit biases can result in discrimination in the workplace and or discriminative behavior against elderly patients. For example, thinking of older people as innocent, cute, adorable, or fragile and in need of protection may seem positive but in fact reflects stereotypes. Infantilizing elderly patients can lead to patronizing behaviors, cause loss of a patient's self-esteem, motivation, and feelings of control and thereby, negatively affect health outcomes.

All of us, whatever our age, arguably need to be aware of the propensity to hold ageist attitudes (which, incidentally, are highly compatible with loving and positive attitudes toward elderly members of one's own family). The editors express wonder that so little concerted effort has been made to counteract ageism, either through advocacy on behalf of older people or through self-advocacy by older people. Ageism, in some way, has been condoned, as witnessed by shelves of noxious greeting cards that perpetuate stereotypes. Physicians, nurses, social workers, and many others in both practice and policy-development roles need to be aware of ageism.

Photo by Dr. Michael Federle

"Getting old is a fascination thing.
The older you get, the older you want to get."

– Keith Richards

Grit and Grizzle

A few years ago, a close friend mentioned that she was headed to play a tennis match.

As my degenerating lower back was starting to limit my long-time love of running, I thought, "Wait, I played tennis in high school, I should get back into it, how hard can it be?" So, I dusted off my 40-year-old racquet and took it into the tennis shop for a fresh set of strings. "You don't want this thing re-strung, sir, you want it framed", I was told. They actually had the same circa 1976 Bjorn Borg racquet up on the wall as a relic of tennis past. So, I picked up a modern, new racquet. Soon after, I challenged that friend to a match. She destroyed me 6-0, 6-0. I blamed the new racquet and tight strings.

Slowly, I started getting back into the sport, and though I was a real hack, I was determined to improve. I felt I was getting good exercise, and it was fun to try something "new". I also started watching more professional tennis matches on TV and instructional videos, studying hitting techniques and strategy. Often these days I am entranced in the Tennis Channel. Anything besides cable news, right?

I am especially inspired when watching some of my favorite young pros like Naomi Osaka (23 years-old) and Stefanos Tsitsipas (22). Their skills, determination and grit are enviable. I often think about my own daughters, slightly older than these players, and how I tried to instill in them the traits of commitment, grit, and kindness, especially when facing adversity or during the long haul of their education. My life tips to them are "notches in the belt of wisdom", I'll say, and at minimum they get a chuckle out of that.

My tennis game has been gradually improving but the ups and downs are humbling. About a year ago I started playing regularly with a long-standing group of guys that play every weekend at Stanford. This is an eclectic group of various current and former professors and other local guys who love the game. Many have been in the group for over 20 years: this is a bunch of grizzled tennis veterans. The average age is probably somewhere in the 70s, with several guys in their 80s (I bring the mean age down a tiny bit). One thing I have learned the hard way is don't let a little grizzle fool you. Never under-estimate your opponent, young or old. These guys may not have Naomi's nimble moves or Stefanos' serve, but they are cunning competitors and they have refined tennis skills and tons of experience. There may be creaking joints, grunts, and groans as they lunge for the ball, but they make contact, get it over the net and keep the ball in play. Their crazy slices and lofty lobs will destroy you.

Reference: https://news.stanford.edu/2016/01/26/herbert-abrams-obit-012616/

These new tennis experiences helped put my "middle" age into perspective. I think how nice it will be (God willing) to be playing into my 80's. One of my favorite radiology colleagues of all time was Herb Abrams M.D, former professor emeritus, and one of the founding fathers of cardiac and peripheral angiography. Herb used to stop by my office, and we'd chat about his tennis matches out "on the vineyard" (aka Martha's Vineyard) where he would spend his summers. On his 95th birthday, he played four generation tennis with his son, grandson, and great-grandson. He played doubles three times a week up until a month before his death in 2016. Herb was incredibly inspiring, the embodiment of grit and grizzle. I wish I would have had a chance to play him.

Like Herb and my new tennis buddies, I am getting a little bit grizzly. I have a few creaking joints of my own and the hairs on my head now number in the single digits. But my game is slowly improving. I am becoming more consistent and I'm trying to take the long view. It is great to have the inspiration of the gritty youngsters as well as the grizzly older players to help frame my viewpoint. In the meantime, I'm almost ready to re-challenge that friend, and I am optimistic that I may actually beat her someday - maybe when I'm 82. In the meantime, I'm working on my kick serve and drop shots.

Christopher F. Beaulieu, MD, PhD

Professor of Radiology, Associate Chair for Education
Stanford Medicine | Radiology

A Journey through Academic Medicine - *Why I will retire on August 2 and return to work on August 3*

In the 40 years since I finished my fellowship I have received many (unsolicited) offers for employment in the private radiology sector, always at salaries so far in excess to what I was making that I was hesitant to discuss them with my wife! However, I have never regretted for a moment the decision to pursue an academic career, and the salary difference ultimately became totally irrelevant. My academic interests have significantly enriched my daily and weekly work activities, and have introduced me to radiologists from around the world who have become collaborators, colleagues and friends. We still see each other on a regular basis and now have grandchildren to compare and brag about as well as our children.

Academic radiology has not only enriched but has extended my career. I am quite sure that the grind of personally reading through lengthy lists of exams and/or performing interventional procedures (which I did for most of my career) would have proved too taxing at my age. Having a steady supply of bright, eager, new residents and fellows, all of whom can decipher new technology faster than I, is a source of relief as well as reward, as my experience and "wisdom" hopefully offset my technical deficiencies.

We in radiology are especially fortunate that our work is not physically as demanding as in some other specialties. Many academic radiologists in their 70s are still welcomed as active contributors by their department chairs. Sam Gambhir has certainly encouraged many of us senior radiologists to remain active in the Department, and has been generous and creative in helping to work out duties and work schedules that match our own evolving needs as well as those of the Department.

I will be "retiring" on August 2 this year, and returning to work August 3 as an Emeritus Professor, working two clinical days per week while continuing my teaching and other academic pursuits. Academic radiology has proved to be a source of great and enduring satisfaction, for which I am eternally grateful.

Michael P. Federle, MD
Professor of Radiology
Stanford University

Dr. Federle is a hobby photographer. An example of one of his photos is shown here and several others are shown throughout this book.

Experience Matters

When I first came to Stanford, I was nearly the same age as the residents. Now, 26 years later, I'm older than many of their parents. We all joke about "senior moments"—forgetting where you put the keys, searching for eyeglasses while you're actually wearing them, etc. Yet it seems that denigrating and mocking our elders is the last permissible form of prejudice in the mass media. I laugh at Saturday Night Live skits like the one showing Amazon's "Alexa" smart-speaker "specially designed for the Greatest Generation" responding in an extra-loud voice to any name the memory-challenged senior happens to call it. But hearing stories about brilliant middle-aged engineers getting laid off from Silicon Valley companies and then unable to find other jobs despite years of experience isn't funny at all. Ageism is rampant, especially in Silicon Valley where people old enough to remember the time when cherry orchards lined El Camino Real in Sunnyvale are viewed as useless by twenty-something tech entrepreneurs expecting to be the next Mark Zuckerberg.

Fortunately things are not as bleak in medicine. The stereotype of the wizened, seen-it-all physician works in favor of older professionals like me. Indeed, it actually worked against me when I was younger and looked more like a high school student than a seasoned pro. While it can be depressing to think that I have more years behind me than ahead of me, my job itself in Radiology is easier than it was when I first started. Having seen hundreds of thousands if not millions of cases over the years, I can draw upon experience instead of memorized lists of differential diagnosis when interpreting studies. And even with all this background, I still see entities every day that I've never seen before, stimulating me to still learn more.

I'm confident that experience benefits not just me personally, but also the patients and trainees I work with. Patients are helped by older radiologists with expertise in both new modalities like CT, which show anatomy and pathology in vivid detail, and older modalities like plain films and barium studies where abnormalities may be subtle and difficult to distinguish from normal variation. Trainees can benefit from teaching in these "lost arts," as well as from the personal connections I've made with radiology colleagues all over the country when they are looking for jobs.

Yet I'd like to think the benefits of having older faculty extend beyond the mere transactional. Belonging to the community of Stanford Radiology, we can all share life's ups and downs—finding significant others, starting families, getting first jobs, losing loved ones—while learning from one another, helping patients and their doctors and getting the work done together.

Terry Desser, MD

Professor
Stanford Medicine | Radiology

"Life is a journey with problems to solve, lessons to learn, but most of all, experiences to enjoy."

– Anonymous

Photo by Dr. Michael Federle

Celebrating aging across cultures: An African perspective

Different cultures have different attitudes and practices around aging and death. These cultural perspectives can have a huge effect on our experience of getting older. I present an observed African perspective focusing mainly on how aging is experienced and celebrated, taking into account the relatively huge gap in life expectancy. People are expected to live to about 78 years in the U.S. and about 60 years in sub-Saharan Africa. People of this age group are still a relatively large work force in the U.S.

Every now and then, I visit the geriatric home where my wife works and it is usually a reminder of how different cultures honor old age. "How often do these people get visits from their children or loved ones?" I asked. "It depends on the race. A few get daily or weekly visits. Most residents do not get frequent visits and some residents have never had a visitor for years," she replies. In the U.S., taking care of the elderly at times can be seen as a burden. But in Africa, it's considered a blessing. In many parts of Africa, it may be considered cruel and abominable to send old and sometimes helpless people into nursing/geriatric homes, particularly when done against their will. Geriatric homes are practically non-existent in most of sub-Saharan Africa. An African widow aged 70 or older, who does not own a home is typically living in a house built by her children with relatives taking care of her. She is always spending quality time with neighbors, children and grandchildren who constantly visit her as a source of wisdom, childcare and knowledge. Children who emigrate from Africa living in the U.S. often invite their parents to visit and are expected to take care of them when they fall sick. The African usually buries their parents with an expensive farewell ceremony. A vigil is held at the family home where the whole community comes to pay respect and offer condolences to the family. Drinks, songs and traditional dances are constant signatures that accompany such an event.

On a number of occasions, in the West, I have seen a teenager grudgingly vacate his or her seat in buses or trains for a more elderly man or woman standing close by only after a conductor's intervention. Other passengers stay unconcerned for the most part, maybe because it is not their child. I grew up in Africa where communities tend to be more intimately bound by cultural and social ties with parents generally feeling a common responsibility to raise all children. Such a concept will be seriously frowned at in the west. It is not uncommon to frequently hear phrases like "Respect your elders." Christian parents constantly quote the Bible to children using choice examples like "Honor your father and mother…" and I am sure across Africa, many children growing up must have heard different versions of the popular African saying "an old man sitting on a stool can see farther than a young man who has climbed a tree." The concept that the village raises a child is emphasized across Africa with messages to respect elders echoed even louder the more elderly the person is. A full crown of white locks on the presumably wiser head invites and commands much respect. The effect of this is that children grow to respect and protect elders. Very often you see several teenagers competing to give up their seats to an elderly person inside a bus, train or church.

Furthermore, in popular culture in the U.S. especially in Hollywood, physical signs of aging are viewed/perceived as hallmarks of a disease and products are constantly advertised on TV, driving the age-sick to purchase and conceal even the slightest wrinkle or gray hair. There is some degree of shamefulness and fear associated with aging, because of the real threat of possible age discrimination, exclusion, poverty and general insecurity. On a public level, similar concerns around aging remain uncommon in Africa and elderly care and concern is not really a pressing policy issue. In fact in many cases, aging is looked forward to with all its attendant perks from the clan.

By 2050 all regions of the world except Africa will have nearly a quarter or more of their populations at ages 60 and above [1]. Currently, the population in the U.S. is aging more than in Africa. Unlike what's been happening in the U.S., where there's a very explicit recognition that an aging population is a very serious developing issue that requires planning and actions, this is not yet the case in Africa. In the U.S., the aging population is very diverse with different groups experiencing aging differently depending on their cultural roots. I believe with the right policies and attitudes in place, life as we age can be more happily embraced, secure and full of opportunity. I believe that older people are one of the most interesting and wonderful cohorts of people on the planet, with plenty that the younger folks should learn from and grow to respect. They need special recognition globally. Doing so would make the world a better place for all of us. *Aging is an inevitable path for us all, a blessing if it was not denied by an early grave.*

Benedict Anchang, PhD

Instructor
Stanford Medicine | Radiology

[1] https://esa.un.org/unpd/wpp/Publications/Files/WPP2017_KeyFindings.pdf
[2] https://www.ted.com/talks/jared_diamond_how_societies_can_grow_old_better

Balancing risk-taking creativity with mindful decision making

I am not quite sure what ageism means but I assume it includes "age discrimination". To clarify my opinion on the topic of "age" in academic medicine, the following ironic illustration that I often used to start a lecture can be used:

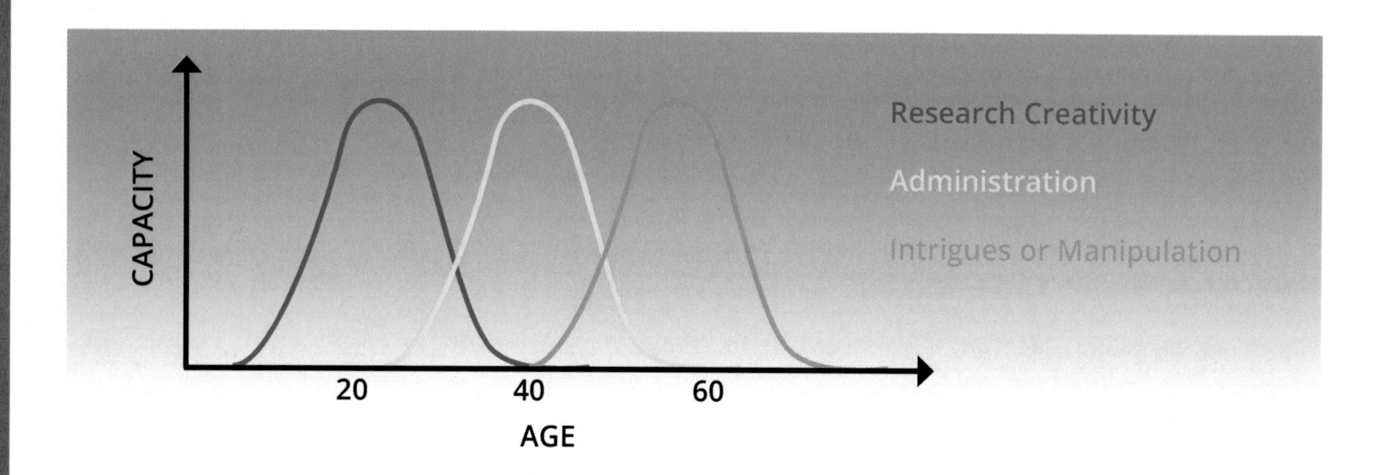

As you can see, the age-axis shows different cycles of activity, with no activity after the common retirement age in Europe, 65 or 67 years. After that, you are not even expected to have any capacity for "intrigues or manipulation". Just sitting on a bench in the park feeding the birds. This is in my mind "age discrimination". And it is, as far as I know, illegal in the US. It is also interesting to compare this graph with a different attitude in many parts of Asia, where to my knowledge there is no age discrimination and age is regarded as a valuable asset.

If I would write a letter to my 30 year old self today, I would ask myself to appreciate my "capacity". It was different 50 years ago: I was thinking faster and had more physical strength. In addition to my work in academic medicine, I always loved woodworking and repairing furniture or even entire homes. However, 50 years ago, I measured or calculated once and completed every task very fast. This often meant that things had to be redone. Today I measure many times and plan more thoroughly. Every task takes much longer, but there is no redoing.

In the American workplace, speed, aggressiveness and action are appreciated qualities. As these traits are also characteristic of young age, there is a risk for indirect discrimination of older people. In general, I think task-groups and project leadership in the US would benefit from a broader age diversity. It is good to balance risk-taking creativity with mindful decision-making. The Asian appreciation of experience, consideration, and poise is easy to understand when you get older.

We are born with a set of traits that are modified with training and experience, but they might not change as much as you might think or wish. An important part of aging is to get to know yourself, your inherent strengths and your weaknesses. If your capacity decreases, you should still say yes to new tasks, analyze your working situation (if you are allowed to keep on working) and balance between your part of the tasks and delegation. An unavoidable aspect of "ageism" is maintaining health. You may become ill, so include that in the equation. In addition, mental health and age related decline in cognitive capacity is difficult to handle. You may not see it coming and you have to rely on friends and coworkers to inform you, if there is a problem. But until then, enjoy working if you wish or feed the pigeons if you prefer that.

Hans Ringertz, MD, PhD

Adjunct Professor, Radiology
Associate Chair, Special Projects
Chair, Nobel Assembly, Karolinska Institute 2002

Some guy said to me: Don't you think you're too old to sing rock 'n roll?

I said: You'd better check with Mick Jagger.

— Cher

You don't have a right to the cards you believe you should have been dealt. You have an obligation to play the hell out of the ones your're holding.

— Anonymous

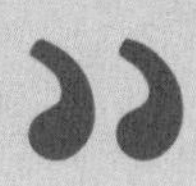

Old age is always fifteen years older than I am.

— Oliver Wendell Holmes

I have reached an age when, if someone tells me to wear socks, I don't have to.

— Albert Einstein

At age 20, we worry about what others think of us. At age 40, we don't care what they think of us. At age 60, we discover they haven't been thinking of us at all.

— *Ann Landers*

Your life's scorecard is all about what happened to the people you led.

— Anonymous

Do You Think You are Too Young?

Think Again!

1) Helen Keller became deaf & blind at the age of 19 months. She was the first deaf blind person to earn a Bachelor of Arts degree

2) Mozart composed music from the age of 5

3) Shirley Temple was 6 when she became a movie star on "Bright Eyes"

4) Anne Frank was 12 years old when she wrote her diary

5) Magnus Carlsen was 13 years old when he became a chess grandmaster

6) Nadia Comăneci was 14 years old when she scored seven perfect 10.0 as a gymnast and won three gold medals at the Olympics

7) Tenzin Gyatso was formally recognized as the 14th Dalai Lama at the age of 15

8) Pele soccer superstar was 17 years old when he won the world cup in 1958 with Brazil

9) Mark Zuckerberg started Facebook when he was 19 years old; Elvis was a Superstar by age 19

10) John Lennon was 20 years & Paul Mcartney was 18 when the Beatles had their first concert in 1961

11) Jesse Owens was 22 years old when he won 4 gold medals at the Olympics in Berlin

12) Beethoven was a Piano virtuoso by age 23

13) Issac Newton wrote the Philosophiæ Naturalis Principia Mathematica at 24 years of age.

14) Roger Bannister was 25 years old when he broke the 4 minute mile record

15) Albert Einstein was 26 years old when he wrote the theory of relativity

16) Lance E Armstrong was 27 years old when he won the tour de France

17) Michelangelo created two of his greatest sculptures "David" and "Pieta" by age 28

18) Alexander the Great had created one of the largest empires of the ancient world age 29

19) J.K.Rowling was 30 years old when she finished the first manuscript for Harry Potter

20) Amelia Earhart became the first woman to fly solo across the Atlantic Ocean at age 31

21) Oprah was 32 years of age when she started her talk show

22) Edmund Hillary was 33 years of age when he became the first man to climb Mount Everest

Do You Think You are Too Old?

Think Again!

23) John Warnock founded Adobe at age 42 and invented the Portable Document Format (PDF) at 50

24) Samuel L. Jackson took interest in drama in his early 20's. But he only became world famous when he was 46, for his outstanding role in Pulp Fiction

25) Suzanne Collins was 46 years old when her trilogy Hunger Games hit the shelves

26) Charles Darwin was 50 years old when he published On the Origin of Species

27) Leo Goodwin was 50 years old when he founded GEICO

28) Bram Stoker wrote his most successful novel, Dracula, when he was 50 years old

29) Ray Kroc was 52 years old when he joined the California Company McDonald's. He facilitated national expansion of the company, which eventually lead to global franchise.

30) John Pemberton was 55 years old when he came up with the famous recipe for Coca-Cola

31) Takichiro Mori was an economics professor. At age 55, he left academia to become a real estate investor, ultimately becoming Forbes' two-time richest man in the world

32) Arnold Schwarzenegger was 56 years old when he became the governor of California

33) Miguel de Cervantes was 58 when he wrote his most famous novel Don Quixote

34) Momofuku Ando developed the instant ramen at the age of 48

35) Judy Dench was 61 years old when she played M in Goldeneye

36) Ricardo Montalbán was 62 when he received the role as Khan in Star Trek II: : The Wrath of Khan

37) Colonel Sanders opened the first Kentucky Fried Chicken franchise when he was 62 years old

38) Betty White was 63 years old when she began The Golden Girls Show

39) A. C. Bhaktivedanta Swami Prabhupada was 69 years old when he started the Hare Krishna movement and founded the International Society for Krishna Consciousness (ISKCON)

40) Mother Teresa received the Nobel Peace Prize for her charitable efforts at the age of 69

41) Anna Mary Robertson Moses was 76 years old, when started painting her first canvas

42) Nelson Mandela, became the first president of post-apartheid South Africa at the age of 76

43) Emmanuelle Riva nominated for Best Actress at the Academy Awards at age 85

44) Gladys Burrill completed the Honolulu Marathon when she was 92 years old. She holds the world record for oldest female to complete a marathon.

"Age is an issue of mind over matter. If you don't mind, it doesn't matter."

Mark Twain

"None are so old as those who have outlived enthusiasm."

Henry David Thoreau

"You are never too old to set another goal or to dream a new dream."

C.S. Lewis

"Let us never know what old age is. Let us know the happiness time brings, not count the years."

Ausonius

DISABILITIES AND ABLEISM

"The most beautiful people we have known are those who have known defeat, known suffering, known struggle, known loss, and have found their way out of the depths. These persons have an appreciation, a sensitivity, and an understanding of life that fills them with compassion, gentleness, and a deep loving concern."

– Elisabeth Kübler-Ross

Disabilities and Ableism

Our next chapter is dedicated to our colleagues, trainees, staff and patients with disabilities. I was thoroughly impressed by the many contributions that we received for this topic. It shows that our community deeply cares about it. Preparing this chapter took us on a journey of many interesting and important discussions through which our team learned a lot:

Anyone of us can become disabled in an instance, through an accident, as the result of a chronic disease, after a surgery or as a result of the aging process. Disability can be visible or invisible, transient or permanent and manifest in many different forms, including but not limited to mobility difficulties, hearing impairment, blindness and intellectual disability, among many others. When referring to disability, some people prefer a people-first language where they are called, "a person with a disability". This style is reflected in major legislation on disability rights such as the Americans with Disability Act. Other people prefer identity-first language which describes the person as "disabled". For example, deaf communities in the United States prefer identity-first language.

For our colleagues who live with disability, just as for everyone else, it is important to be recognized as unique and capable individuals. While every person is different, many people with disability have developed an amazing patience through their life experiences of waiting for almost everything. Many people with disabilities are masters in perseverance, overcoming obstacles, and not stressing out about the little things in life.

Ableism describes the process of discriminating against people with disabilities and favoring able-bodied people. At Stanford, we want to see through perceived or actual imperfections, so that everyone's unique light can shine through. We believe that it is our collective duty to enable our colleagues with disabilities to express, contribute and realize their unique talents and ideas. This chapter shows how people with disability shape our collective experience and how their important insights help us creating a better future for all of us!

How to be an ally

In medicine, our practice has always involved people with disabilities. However, biases and stereotypes of abled-bodied healthcare workers on one side and physically/psychologically impaired patients on the other side can lead to poor integration of healthcare workers with disabilities into medical care teams, poor patient care and negative health outcomes. The points below provide examples, how to be an ally for team members or patients with disability. These points are not comprehensive and may require adjustments to specific situations:

- Educate yourself and your community about different models of disability.
- Hire and promote students, staff and faculty with disabilities
- Explicitly communicate employment equity in your job advertisements and workplace policies. Outline clear anti-discrimination and harassment policies.
- Ensure accessibility of buildings, workplaces, patient examination rooms, restrooms, cafeterias, conference rooms, etc. and any place where social gatherings occur
- Offer team members with disability to submit a Disability Impact Statement, which assists staff by outlining the individual's impairment and any modifications they may need.
- If you are able-bodied, do not assume that the quality of life experienced by people with disabilities is lower than your own.
- Invest extra time in communicating with people with disability. Listen. Understand their specific needs and talents. Ask open questions and do not assume that you are entitled to learn all about a person's disability, just because you noticed they are disabled.
- Do not discuss someones disability in front of others without their permission.
- People with disabilities can flourish in a diverse range of bodies and in different ways. Focus on what someone can do, rather than what they cannot do.
- If you work in a large organization, encourage and support the formation of a network or resource group for team members with disabilities.
- Do not take or exploit financial resources, employment opportunities, or positions of power in diversity-related fields from people with disabilities.
- If you are applying for research grants or other funding, direct the acquired funds back to the community, e.g. by hiring researchers, assistants and students with disabilities.
- Recognize bias and discrimination against people with disabilities. Speak up if you notice intentional or unintentional microaggressions.
- Provide easy access to sign language interpreters for people with hearing disabilities
- Face deaf people when you talk with them, avoid looking down or covering your mouth while speaking, provide videos with captions, distribute meeting minutes.
- When working with visually impaired people, verbalize as much visual information (e.g. while writing on a white board or showing a powerpoint file) as practical and feasible. Record lectures and make all materials available online for advanced postprocessing.
- Establish partnerships with advocates and stakeholders before making any decisions, initiating research, creating policies or planning workshops on diversity-related topics.
- Make sure your website, blog, article, podcast, video or social media post is accessible.
- Follow advocates for people with disability on social media. Read and engage with content creators and influencers. Amplify their voices by sharing their work.
- Purchase from businesses led by people with disability
- Do not assume that this list does not apply to you.

I suffered a spinal cord injury in 2003 during my gastroenterology fellowship at UC San Francisco. Unable to practice independently or perform procedures, I decided to pursue radiology, completing residency and fellowship at Stanford. Interviewing for residency in my wheelchair, attendant in tow, was very intimidating, and not all programs were enthusiastic about the challenge of training a partially quadriplegic person.

Fortunately, Stanford was committed to customizing the residency around my abilities. I didn't want to do less work than others, but it was clear that it would have to be structured differently. Thinking creatively, some solutions included modifying my call schedule to shorter but more frequent shifts or assigning me to work up consults rather than performing interventional radiology procedures. And similar modifications continue. I am grateful to work with such incredibly understanding and supportive colleagues.

Although I can interpret images and dictate without assistance, there remain a myriad of unseen challenges. From morning until evening, I have paid caregivers and hospital volunteers who help me with all of the mundane tasks that I used to take for granted. This includes getting ready for work, grabbing a coffee, and helping me onto my Segway. But I feel like literally everyone helps me out, from moving a chair out of my way, opening a door, to finding me a straw. These infinite small kindnesses get me through the day.

Peter Poullos, MD

Clinical Associate Professor
Stanford Medicine | Radiology

Dr. Peter Poullos presenting to attendees in the SMAC launch event

Making an Invisible Community Visible:

Founding the Stanford Medicine Abilities Coalition (SMAC)

People with disabilities are the largest minority in America. Yet, regardless of whether the disability is visible or invisible, people with disabilities are often invisible in discussions about diversity and underrepresented in the field of medicine. Disabilities are not just those that you can see. In fact, there are a variety of disabilities, including physical, hearing, visual, psychological and learning. Moreover, two people may share the same disability but be impacted quite differently and have vastly different experiences. Those experiences are shaped by their communities, both inside and outside of work. In the workplace, the person with the disability may encounter barriers and supports, the balance and nature of which depends on the institution's structure, as well as on culture and climate. Supportive structural arrangements include access to appropriate accommodations, ease of access to those accommodations, knowledgeable disability service providers, and rich personal networks with dedicated student and faculty organizations. Cultural and climatic attributes that can improve the lives of those with disabilities focus on ongoing professional development, awareness, and openness [1]. Specific considerations may include the regular assessment of institutional policies, services, and space; the incorporation of disability into diversity and inclusion initiatives; openness to disability in admissions and recruitment; and deep leadership commitment to diversity and inclusion, among others.

In the fall quarter of this academic year, Dr. Peter Poullos, a Body Imaging Radiologist at Stanford joined the School of Medicine's Faculty Senate Subcommittee on Diversity, led by Dr. Iris Gibbs. The subcommittee realized that disability was one of the underrepresented groups at the medical school without representation or affinity groups.. Thus, the subcommittee decided to focus on disability as their lead project for 2019. Shortly after, Dr. Poullos attended the "Future of Disability at Stanford" event sponsored by the Stanford Disability Initiative, where he learned about campus-wide initiatives aiming to incorporate disability as a fundamental part of diversity, inclusion, and equity. Through these two separate organizations, he recognized that the School of Medicine needs an organization to promote community, support, and advocate for students, trainees, faculty and staff with disabilities.

To fill this void, he founded the Stanford Medicine Abilities Coalition (SMAC).. SMAC is composed of people who have disabilities and their allies. SMAC aims to foster and advocate for the equal treatment and well-being of everyone at Stanford Medicine, regardless of differences; to promote collaboration between people who are passionate about disability at Stanford Medicine; and to advocate for accessibility, resources and disability services at Stanford Medicine, above and beyond that required by law. The organization does not enforce legal requirements, rather the focus is towards actively welcoming and even recruiting people with disabilities. This is a fundamental shift in emphasis away from merely making "reasonable accommodations," to realizing that those with disabilities have a unique perspective on the healthcare system and on life in general, which is necessary to include in an organization devoted towards caring for others. Who better to understand our patients than those who are both healthcare providers and healthcare recipients?

Dr. Poullos has had a disability since 2003 when a bicycle accident caused a spinal cord injury. No longer able to perform endoscopic procedures or practice independently, he came to Stanford in 2004 to start over with a second residency and fellowship in radiology. He joined the faculty in 2009, at which time he became Associate Radiology Residency Program Director, a role he performed until 2016. Having transitioned successfully into his new field, he turned outward towards diversity advocacy and later this disability work which led to the formation of SMAC. Having not worked with disability groups at Stanford, he had much to learn. Luckily, he was able to connect with Zina Jawadi, the chair of the Stanford Disability Initiative, and with Richie Sapp and his fellow medical students in the Medical Students with Disability and Chronic Illness (MSDCI) group. They educated him on disability organizations and issues at the University and at the School of Medicine. He turned to unraveling the alphabet soup of disability organizations, including the DAO, OAE, SLC, OFDD, and the COE (Diversity & Access Office, Office of Accessible Education, Schwab Learning Center, Office of Faculty Development and Diversity, and Center of Excellence in Diversity in Medical Education). In March, with sponsorship from the Dean's Office and the OFDD, SMAC representatives joined the National Coalition for Disability Access in Health Science Education and flew to Washington D.C. to attend their Annual Symposium. There, they were able to connect with national leaders in this field.

SMAC's Accomplishments

SMAC held its launch event in April, an informational session and mixer with MSDCI. Over 70 guests, including Dean Lloyd Minor and the Chair of Radiology, Dr. Sam Gambhir came together to engage in critical conversation about disablity inclusion. The event brought together a diverse group of people from the School of Medicine, united in the common cause. SMAC held its first dinner and brainstorming meeting in May. We hosted a booth at the Second Annual Diversity and Inclusion Forum and Fair. Most importantly, SMAC now has a seat in the School of Medicine Diversity Cabinet, only the second "identity group" after the LGBTQ+ /SGM to be represented in this fashion. Still, we are at the early stages of forming the organization, currently building organizational leadership, social media presence, and outreach. SMAC is active on social media with a Facebook group, Facebook page, Instagram page, Twitter account, and a YouTube channel. On a practical level, we have been working with the Clark Center management to make the Peet's Coffee and classroom area more accessible for people with physical disabilities.

Making an Invisible Community Visible:

Founding the Stanford Medicine Abilities Coalition (SMAC)

Moving forward, SMAC will host quarterly events and monthly meetings. So far, we have three committees who will focus on supporting disability service providers in medical education, address issues related to residency and fellow-ship, and fostering an inclusive institutional culture for faculty and administration. We will continue to serve as mentors for MSDCI. Dr. Poullos will be participating in a panel discussion on diversity at the Department of Radiology Diversity Fair in September. Lastly, we are currently putting together a survey about the state of disability at Stanford School of Medicine, set to be launched in October.

We hope for our work to motivate individuals with disabilities and their allies to become involved with our initiatives. At the same time, we would like to highlight that society often portrays people with disabilities in the lens of inspiration simply for completing the same daily routines as someone without a disability. We want to avoid this pitfall, seeking to inspire but not be inspirations or inspirational per se. The distinction is small but important. We are extremely excited about the future of SMAC, because it has the potential to transform the lives of the largest minority in Health science. Most importantly, we hope to make disability an integral part of diversity and make visible an otherwise invisible community.

Peter Poullos, MD

Clinical Associate Professor
Stanford Medicine | Radiology, Gastroenterology and Hepatology
Founder and Director
Stanford Medicine Abilities Coalition (SMAC)

Zina Jawadi, MS

MS Candidate
BS, Biology
Stanford Medicine | Bioengineering

Sources:

1. Lisa Meeks and Neera Jain, Accessibility, Inclusion, and Action in Medical Education: Lived Experiences of Learners and Physicians with Disabilities, AAMC March 2018

https://www.cnn.com/2015/05/20/health/human-factor-philip-zazove/index.html
https://news.aamc.org/diversity/article/paving-way-med-students-physicians-disabilities/

https://youtu.be/SxrS7-I_sMQ

We would like to thank Dr. Lisa Meeks, an Assistant Professor at the University of Michigan and an expert in Disability in Medicine, for reviewing this article

An Interview of Dr. Payam Massaband by Dr. Heike Daldrup-Link

Dr. Payam Massaband has been a member of the Stanford Department of Radiology since 2005. He completed fellowship in cardiovascular and musculo-skeletal imaging and currently serves as Chief of the Radiology Service at the VA Hospital and Program Director for the Stanford Radiology Residency Program. I had the pleasure talking with him about his insights for our Radiology Department and reflections on disability:

What would you like us to know about yourself?

I would like our trainees and colleagues to know that I appreciate honest communication. That's what I like about the diversity newsletter and that's why I proposed this interview. Our Stanford community values kindness above all else, which is great. But for some this has meant never saying anything critical or, God forbid, having difficult conversations. I don't see that kindness and honesty need to be mutually exclusive. I also think that it is important for us to understand the diverse communication styles that exist. Diversity in communication, as long as it is respectful, will enrich our community.

Professionally, I completed a BS in Neuroscience at UCLA in 1998 and an MD degree at USC in 2002. After that, I started a residency in the Department of surgery at Stanford. In 2005, I switched to radiology. In my third year of surgery residency, I had a few episodes of falling that I attributed to my shoes. Despite throwing away those otherwise fine shoes a colleague asked why I was limping. "Just some hip pain" I answered. I suppose this is the definition of denial but, in the moment, I had no idea whatsoever. After I discovered foot drop on a quick self-exam I realized there was something wrong. I got a spine MRI hoping for a benign tumor! Over the ensuing months I diagnosed myself with a very rare myopathy. Given this new diagnosis, I decided to switch my training. I considered psychiatry and radiology and after deliberating pros and cons, I decided on radiology. In an extreme twist of fate these days I mostly practice psychiatry.

How did interactions with your colleagues change when you discovered your disability?

My surgical colleagues were unbelievably supportive. In radiology there were some who questioned whether I could physically perform the duties of radiology residency, including call and procedure-related rotations. One recommended that I should go become a consultant. At this point it is important for me to mention Dr. Brooke Jeffrey, who was absolutely wonderful to me throughout this very difficult time (and since!). When it became clear that I had to leave surgery, I called Brooke at his home late one evening. He was a most kind, gracious and supportive leader. He sensed the anxiety and fear in my voice and simply said: "Do not worry Payam, you will be fine." I have learned so much from Brooke, and others who came to my side at that time, including my surgery program director Dr. Ralph Greco and chair Dr. Tom Krummel, and try to pattern my leadership after their example.

Where can we as radiology colleagues do a better job to accommodate people with disability?

Accessibility is an important issue. Before I plan to attend any meeting or conference, I have to make sure I can get there. Stanford has some areas where accommodations are great and some opportunities for improvement. There can also be challenges at gatherings outside of the Stanford campus, for example at a faculty dinner at a restaurant or someone's home. Stairs tend to be an issue since most people don't even notice them so they forget to mention them ahead of time. I should say that I am able to make it through the day due to the innumerable acts of kindness and support I get everywhere I go.

What was a surprise to you with regards to disability?

I would never have believed this before I experienced it repeatedly: sometimes people do not seem to see you. It is a very odd experience when you are with a group and someone joins and never looks at you or talks with you. You can be in a bar and try to make your way through and nobody moves an inch. Or you can be in a waiting line and people just cut in in front of you. I don't think this is any bad intention. I believe that some people just literally do not see a person in a wheelchair.

Do we get applications from students with disability in your current residency program?

Medical students with disability are uncommon. Medical schools have historically either discouraged or not made clear that they will accommodate people with disabilities. The job demands are often so cumbersome and unaccommodating, that I suspect students with disability often choose other areas.

While we are discussing applications for our diversity newsletter, I want to let everyone know that I am very sensitive to diversity in our program. We are still relatively lacking in female and URM residents. We have many efforts, both within the selection committee and in collaboration with the School of Medicine and University, geared at improving the recruitment of a diverse group. I also have a very broad view of diversity, which includes diversity of experiences and goals. Our strategy is to work hard to recruit truly outstanding residents. Women, men, URM, PhD, DO, research-minded, LGBTQ, first-to-college-in-the-family, entrepreneurial, private-practice-bound, etc. Broad diversity enriches us all.

What will come next for the residency program?

I think it is amazing that we still follow teaching concepts that are decades old. Our Residency Program Committee is working hard to create a 21st century learning experience: For example, we will soon pilot a rotation with a programmed-learning curriculum, where the clinical work of the day is split with a curriculum including lectures and teaching file cases chosen as part of a deliberate teaching goal. Moving forward, we would also like to include advanced computer technologies and artificial intelligence in our training program. Similar to a flight simulator, we could simulate different clinical scenarios for our residents, with programmed interruptions such as contrast agent reactions, which can also be simulated. We could teach, optimize, and test trainee development of expertise in a virtual environment, for example.

What is your biggest strength?

I highly enjoy teaching and I enjoy my administrative roles. On a personal level, I am the youngest of five sons and my huge family all live in Los Angeles. This large family and specifically having four older brothers forced me to learn how to get along with all types of people and communication styles. I am blessed to have a wonderful family of my own now, and that learning continues in my own home: my wife and I have a son, Noah, who is ten and a daughter, Sasha, who is eight.

What does empathy mean for you?

Empathy is the ability to imagine and simulate the minds of others in order to understand how they think or feel. We often talk about putting ourselves in others' shoes or seeing something from someone else's point of view. However, in my mind there is an evolution to empathy: Empathy 1.0 is simply transporting yourself into another person's situation, taking your own biases and life experiences with you. Empathy 2.0 might mean that in addition to putting yourself in someone else's situation, you also try to understand what brought them to this place, their experiences, biases, fears, and limitations. Don't get me wrong, any kind of empathy is encouraged but Empathy 2.0 comes with much less judgment, much more understanding, and a greater sense of acceptance. Maybe I'm just advocating for more compassion.

What life lesson would you want to tell a trainee, with or without disability?

I would want to tell everyone the same thing: Life is full of challenges. Specifically, residency can be very rough. And a training experience in radiology, whether clinical or research, is a marathon. And like the recent Boston Marathon, you will occasionally battle 30 mph headwinds in freezing, slushy weather. You will feel exhausted and fall down. I deal with a very obvious physical disability but everyone has some particular challenges, many less obvious. With the knowledge that you are surrounded by supportive colleagues, you will need to summon all your inner strength and grit to work through these challenges. You will make it, we will do so together.

Make somebody's day today

A hero is an ordinary individual who finds the strength to persevere and endure in spite of overwhelming obstacles.

— Christopher Reeve

My advice to other disabled people would be, concentrate on things your disability doesn't prevent you doing well, and don't regret the things it interferes with. Don't be disabled in spirit.

— Stephen Hawking

Life is all about balance. Since I have only one leg, I understand that well.

— Sandy Fussell

Aerodynamically the bumblebee shouldn't be able to fl , but the bumblebee doesn't know that so it goes on flying anywa .

— Mary Kay Ash

When you hear the word 'disabled,' people immediately think about people who can't walk or talk or do everything that people take for granted. Now, I take nothing for granted. But I find the eal disability is people who can't find joy in life

— Teri Garr

Just because a man lacks the use of his eyes doesn't mean he lacks vision.

— Stevie Wonder

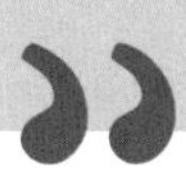

Embracing Indivdual Differences

As a female and first-generation Guyanese-Canadian, diversity is a topic close to my heart.

Workplace diversity encourages broad ranges of talents and perspectives, and can improve both company productivity and morale. I feel incredibly lucky to have been immersed in environments that promoted diversity from a young age. However, I realize that for many others in the medical community and elsewhere, this is not a shared experience. As a current member of the Stanford University and Organization for Human Brain Mapping (OHBM) communities, it is highly encouraging to see diversity committees dedicated to improving the experiences of under-represented groups and minorities. Themes such as "Celebrating Diversity" have been prominent at previous OHBM conferences and happily embraced by my colleagues and I (as displayed on our name badges), among thousands of other meeting attendees.

While diversity is often discussed in terms of ethnicity, gender, religion, and age, it is important to also support the concept of neurodiversity. Neurodiversity embraces neurological differences as part of normal variations in humans. It is a concept that challenges views that assign "disorder" labels to individuals who have received diagnoses such as autism or dyslexia, and helps to break down societal barriers that prevent their inclusion. In recent years, improved awareness about neurodiverse workforces has inspired a number of large corporations to foster environments that better support different perspectives and ways of thinking. This movement has proved to be beneficial for all parties involved.

In general, diversity is a requisite for growth and innovation. As explained by entrepreneur Nick Walker in a recent Fast Company article (Nerenberg, 2017), "In any given sphere of society, we only get the benefit of the contributions of those individuals who are empowered to participate."

Danielle DeSouza, PhD

Instructor
Stanford Medicine | Neurology & Neurological Sciences

The Stanford Neurodiversity Project: Empowering the Neurodiverse Population

In April 2017, several Stanford faculty and I attended the second Annual Autism at Work Summit at SAP. We heard stories after stories from SAP employees with autism about how their jobs have changed their lives. Due to the intrinsic challenges in social interactions, individuals with autism have difficulties maneuvering the typical job search and interview processes. In fact, over 80% of adults with autism are unemployed or under-employed. After we returned to Stanford, a few of us got together and started talking about how we could help with hiring individuals with autism to work at Stanford.

In Fall 2017, I founded the Special Interest Group for Neurodiversity (SIG-Neurodiversity). We define neurodiversity as a concept that regards individuals with differences in brain function and behavioral traits as part of normal variation in the human population. We started inviting speakers to come to our monthly meetings and talk about topics on neurodiversity. In our first meeting in November 2017, we thought we would have only a handful of us, but we had 23 attendees in the first meeting and built a rapidly growing community from there. We have been fortunate to be able to invite very good speakers. The topics usually relate to employment, education, or mental health for neurodiverse individuals. Soon after the first couple of meetings, companies around Stanford started requesting for opportunities to participate in our monthly meetings as well.

In April 2018, Dr. Laura Roberts, Chair of Department of Psychiatry and Behavioral Sciences, suggested to me that I should escalate my efforts on neurodiversity from a special interest group to a special initiative of the psychiatry department. I was appointed the Director of the Stanford Neurodiversity Project (SNP). There are six major objectives of the SNP:

- Establish a culture that treasures the strengths of neurodiverse individual
- Empower neurodiverse individuals to build their identity and enhance their long-term skills of daily living throughout the lifespan
- Attract talented neurodiverse individuals to study and work at Stanford
- Train talented individuals to serve the neurodiverse population
- Disseminate the Stanford Neurodiversity Model locally, nationally, and internationally
- Maximize the potential of neurodiversity

Over the past year, our project has received tremendous support internally and externally. With a major philanthropic support, we are able to grow the project. Currently, we have three main initiatives:

1. Neurodiversity Awareness and Education Initiative
2. Neurodiversity at Work and Wellness Initiative
3. Neurodiversity Independent Living Skills and Housing Initiative

Within the Neurodiversity Awareness and Education Initiative, we have 5 current activities:

- SIG-Neurodiversity. Since its inception, over 240 individuals have asked us to include them in the SIG-Neurodiversity. Participants call in through videoconference from all over the world.
- Neurodiversity Awareness Program. I have been speaking about the strengths-based model of neurodiversity

and the SNP in various departments at Stanford, locally in the Bay Area, nationally (Nashville, Seattle), and internationally (United Nations, Moscow). I will be one of the keynote speakers at the Autism at Work Summit at Melbourne, Australia later this month.

- Courses on Neurodiversity. I started teaching two courses on neurodiversity. The first course, "Topics in Neurodiversity: Introduction and Advocacy" (PSYC223) was offered in Winter 2019. In addition to behavioral, biological, and social perspectives of neurodiverse conditions, we taught Stanford students on how to work with neurodiverse high-school students. In Spring 2019, I collaborated with Dr. Nicole Ofiesh (Director of the Schwab Learning Center at Stanford) on teaching "Topics in Neurodiversity: Design Thinking Approaches" (PSYC223B). We taught students about design thinking approaches, strengths-based model of neurodiversity, and universal design. In this past quarter, we had a medical student, graduate students from the Graduate School of Business, Graduate School of Education, Department of Sociology, and undergraduate students in Computer Science, Human Biology, and Engineering. The 14 students formulated three projects. The first one was on teaching neurodiverse students how to navigate the process of getting accommodations; the second topic was related to preventive mental health materials that can be taught to residence assistants; the third topic was on performance evaluation for neurodiverse employees. PSYC223B will be offered again in Winter 2020. PSYC223 will be reorganized into three 1-unit courses to be taught in Fall 2019, Winter 2020, and Spring 2020.
- Stanford Neurodiverse Student Support Program (NSSP). The NSSP will be launched in September 2019. This program is a comprehensive program designed to enhance the experience of neurodiverse Stanford students in college by providing support on transitioning to college, social life, independent living, accommodations, learning, and mental health. We have already recruited peer mentors who will serve to facilitate the support on transitioning to college, social life, and independent living. SNP staff members will collaborate with existing groups within Student Affairs to provide support on accommodations, learning, and mental health.
- Neurodiversity seminar series for managers and teams working with neurodiverse employees. This seminar series started in February 2019 after a neurodiverse employee was hired in the School of Medicine. After onboarding of the new neurodiverse employee, I provided support to the manager and team through discussion on best practices for working with neurodiverse individuals.

Our second initiative is the Neurodiversity at Work and Wellness Initiative. Please see below for the two active activities for this initiative:

- Neurodiversity at Work Program. This program provides social skills support, workplace training, short-term coaching as well as customized technical training for neurodiverse participants who would like to be considered for open positions. We provide neurodiversity awareness training, position consultation, neurodiversity best practice training, and on-going support to potential employers at Stanford. We have successfully facilitated the hiring and provided on-going support for one neurodiverse individual in the School of Medicine earlier this year. This work in other departments in the University is in progress.
- Adult Neurodevelopment Clinic. This clinic provides mental health support for neurodiverse students and employees. Services include diagnostic evaluations and on-going treatments such as medication management and psychotherapy.

Our third initiative is the Neurodiversity Independent Living Skills and Housing Initiative. The only currently active program is the Independent Living Skills Program. Our group recently started a formal collaboration with Autism Speaks to co-develop materials for teaching service providers and family members on how to work with neurodiverse individuals.

The SNP has been growing rapidly in the past year. For more up-to-date information about our programs, please visit our website (http://med.stanford.edu/neurodiversity.html) or contact us at stanfordneurodiversityproject@stanford.edu.

The Stanford Neurodiversity Project: Empowering the Neurodiverse Population

In this newsletter, I would like to invite you to participate in this movement on neurodiversity. If you like to join SIG-Neurodiversity, please send us an email (see above). Please let us know your interests in neurodiversity. Let us know if neurodiverse individuals you know are looking for employment. Please introduce us to organizations interested in learning more about neurodiversity, and those interested in hiring neurodiverse individuals. Tell us how you want to be involved.

By working together, we can make a difference to the neurodiverse community and beyond.

Lawrence Fung, MD, PhD

Director, Stanford Neurodiversity Project
Director, Adult Neurodevelopment Clinic
Clinical Assistant Professor
Stanford Medicine | Psychiatry & Behavioral Sciences

"Disability is a matter of perception. If you can do just one thing well, you're needed by someone."

Martina Navratilova

*" Start by doing what's necessary;
then do what's possible;
and suddenly you are doing the impossible."*

Francis of Assisi

" Your success and happiness lies in you. Resolve to keep happy. and your joy and you shall form an invincible host against difficulties."

Helen Keller

" Disability is natural. We must stop believing that disabilities keep a person from doing something. Because that's not true . . . Having a disability doesn't stop me from doing anything.."

Benjamin Snow, Grade 8, in his essay "Attitudes About People with Disabilities"

Notes

LIBERALS & CONSERVATIVES

"The world is a dangerous place, not because of those who do evil, but because of those who look on and do nothing."

– Albert Einstein

Liberals and Conservatives

This chapter is dedicated to our liberal and conservative team members. We will learn about actual or perceived differences between liberals and conservatives, what we have in common and how either view might affect our community and our academic environment. How does it feel to be a minority of either side in a community of "the others"? What do family members feel or think?

The First Amendment of the Constitution guarantees our rights to free speech, freedom of the press, freedom of assembly, petitioning the government for a redress of grievances, and freedom to exercise our religious beliefs. Discrimination based on political views happens when someone is treated differently because of their political beliefs, party affiliation, or civic activities. Employers are allowed to limit political activity at the workplace. While related laws are different in different states, in general, employers should avoid treating employees differently based on their political affiliation.

Liberals and conservatives in the United States disagree about fundamental values that should govern our lives. Perhaps we should not only ask, what this divide is, but why we have it. Some of our authors describe that it might be rooted in perceived or actual physical, intellectual or economic disparities. Some people feel left behind. How can we address these concerns?

On one hand, as a community, we have to define what is tolerated and what is not tolerated. We have to communicate our basic rules and defend them if needed. At the same time, within a framework of common values, both liberal and conservative team members might have important insights to offer. Perhaps we can try to judge a little less and listen a little more.

People want to know who we are and what we stand for. Where do we stand in Stanford Medicine? We believe in inclusive excellence. We seek many different head codes to crack the greatest enigmas of our time. We believe that people with talent, purpose and determination can change the world. We want to unite and focus our talents to save lives and nurture our collective humanity.

How to be an ally

How do we create an environment where we do not only listen, but truly hear each other? How can we reach someone who is determined to not understand? The points below provide ideas for tangible action points. These might not work for everyone. The creative reader is invited to expand on these ideas and adjust them to specific situations:

- Understand your own moral foundations and how these could affect your emotional responses. Through effective self-control, communication experts can achieve cooperation with people whose core convictions differ from their own.
- Learn about the history of the community you are serving
- Socialize across the political aisle and engage in frequent informal discussions with people from substantially different backgrounds and viewpoints from your own. Conducting mutually respectful conversations is a skill that needs to be practiced and grown over time.
- Try to find common ground first, such as love for family members, hobbies, professional interests, work experience, etc.
- Do not make assumptions about someone based on their vote.
- Do not start a conversation with the agenda to bring the other person on your side of the political ile. Start a conversation from a point of curiosity. Try to understand.
- Let the other person know that you heard them, even if you disagree. Use the script: "I understand you see it that way.... I see it differently ..."
- If you want to achieve empathy for your point of view, apply the technique of moral reframing: For example, if you want to convince a conservative about the value of national health insurance, connect your arguments with conservative values, such as patriotism and moral purity. If you want to convince a liberal on military spending, connect your arguments with liberal moral values, such as equality and fairness.
- Hire and appoint team members and leaders who represent ideological diversity, viewpoints from across the political spectrum as well as different races, origins, and backgrounds. Create diverse teams who work together on common professional goals.
- Clearly communicate non-negotiable values and policies for all team members
- Communicate that within the framework of common values and policies, all team members are open to exercise their voice and opinion regardless of preconceived notions or political affiliations
- If someone makes a comment that violates your values, speak up!
- Avoid getting into a back-and-forth argument about someones views that are unacceptable to you. Use the following script: "Your comments are hurtful to me. Please do not make these remarks in front of me again."
- Organize discussion groups of liberals and conservatives, where participants are asked to engage in respectfully considering one another's points of view on a specific topic.
- Unless the discussion partner has truly toxic point of views, the goal of these discussions is to listen and understand, getting your point across and maintaining the relationship.
- Discussions about political beliefs are tied to our moral values and can be emotionally overwhelming. Sometimes, it is okay to take a break or just walk away.

We The People

Life can be difficult these days. Some people in our country lost their kindness and say hateful things about "the others." The stuff that some people say and write is terrible and hurtful. I think we have to define what is tolerated: We do not tolerate abusive, hateful or discriminating behavior.

Fortunately, Stanford is trying to create an inclusive environment for everyone. I am grateful for the opportunities that I have been provided at Stanford Radiology. Unfortunately, there are people out there who seem to be unhappy that people like me work at an elite University. Imagine how that feels. I don't want anybody to hate or dislike me. And it would not be accurate to say I don't care. But I don't care enough to change what I am doing. Because what I do is important. I carry on and focus on my goal to help my patients, students and co-workers every day.

What would I want to tell other members of minority groups—or any of my colleagues? I learned this inspiring advice: Whether you have a good day or a bad day, keep your heads up! And there is only one way your heads can be up—knowing that you have given your absolute best out there. When you truly accept this principle, it changes everything: your preparation and performance, your ability to withstand setbacks and overcome obstacles—and it dramatically improves your ability to succeed.

Whether you think people hate each other or care for each other, you will probably be right. Is that due to a difference in the people you meet or what you see in these people?

Anonymous

"There comes a time in every life when the past recedes and the future opens. It's that moment when you turn to face the unknown. Some will turn back to what they already know. Some will walk straight ahead into uncertainty. I can't tell you which one is right. But I can tell you which one is more fun."
PHILIP H. KNIGHT
MBA 1962

Instead of judging people, try to understand them instead

I grew up on the east coast. My mom was, by most definitions, a hippie, who shopped at Whole Foods, then called "Fresh Fields," did yoga, and brought home local, glass-bottled, non-growth hormone milk before any of that was trendy. She voted purely on environmental issues. My dad, on the other hand, was a die hard, free market, Reagonomics fan, and he usually voted at the expense of the whales and other endangered species she was so passionate about. So, at an early age, I was exposed to both sides of the political aisle and heated debate over various organic dinners.

Our neighborhood was affluent, and non-inclusive, a gated community you had to apply to, to be allowed to purchase a home there, and it was 100% white except for one adopted child I can think of. But I went to the local public school, where numbers-wise, I was a definite minority by a long shot. Talking to my parents about that recently, my dad reminded me that my sister and I were the only ones from our neighborhood at our bus stop every morning, something that, as a kid, I don't remember ever thinking about. Everyone else from the neighborhood went to private school. He purports that he wanted us to have "expanded life experiences"… I think part of it was that he was happy it was free—nonetheless, high school was a mixing pot of culture, race, and diversity in which I was totally immersed and constantly learning. I learned about freshman friends getting pregnant, friends without money for lunch, ghettos, welfare, weaves, fights, a love for 90s rap music (which my mom despised), and every now and then, I picked up a hint of calculus.

In the middle of high school, my dad retired and we all moved down south. My parents now say it was because they wanted us to experience "the other side of America," but I think a lot of it had to do with my mom's love for manatees, beaches, and warm water. It was definitely a different side and a culture shock. Though a beach town, it was true Old Florida. Economically, the town had been driven by lemon production, which dried up years ago, and now the blue collar jobs were in fishery or tourism. My first day of school there, I pulled into the parking lot and saw Confederate flags on the back of students' pick up trucks! I was flabbergasted--were those even legal anymore?? I had no idea. I'd never seen one in real life. There was one single African American family at the school. And everyone else was white. It almost felt like a different country. But a lot of the problems were similar to my previous high school—teenage pregnancies, fights, welfare, but I was exposed to new things too—most relevant to politics today, the hopelessness and confusion of feeling left behind without opportunity for progression.

From my experiences at home, to my very different two high schools, I've been exposed to both sides of the political aisle. I'm not going to tell you who I voted for, but I do believe that diversity is not just about check boxes, it's about bringing together people with varied experiences, which may be significantly influenced by their race, gender, ethnicity, religion, sexual orientation, socioeconomic status, and the list goes on. As I learned over my own dinner table, political leaning seems to be almost as engendered in us as the aforementioned traits (though my dad, to everyone's surprise, switched sides in 2008). Nonetheless, instead of judging people by their political leaning, or trying to convince them to switch sides, maybe we should try to understand them instead. Figure out their background, their family life, their path, their inspiration, their worries. Because if we judge someone purely based on political leaning, aren't we being just as bad as the "other side"?

Mary Ellen Koran

Resident
Stanford Medicine | Radiology

We Can Learn From Each Other

I would like to start by clarifying that I do not hold any strong opinions with regards to politics. When Dr. Daldrup-Link approached me to share my opinion, my immediate response was that I do not have anything to share. But she convinced me that everyone has a story. So, here is mine:

My family lives in the Midwest. My sister recently moved to a small town that you probably never heard of. I googled every possible spelling of it and got nothing. That's how small that place it. I know you feel sorry for her right now. Well, she feels sorry for us because we are spending years of our lives in windowless compounds.

It hurts my feelings that Californians seem to think that everyone in the Midwest is overweight, poor and dumb. My sister might be short in cash, but she could outsmart anyone I met in California thus far. For example, when my laptop broke down a few weeks ago, my colleagues suggested I should buy a new one. My sister fixed it. So much for stereotypes. In her world, resources are scarce and she has to be creative to solve problems. We all could learn a lot from her.

Many Californians think they are liberal. But I found a very tight culture here: In California, you have to dress in a certain way, drive certain cars, eat the right food and speak in a certain way. Any deviation from these norms is socially disapproved. If I had my druthers, I'd have a Chocolate Eclair with my healthy lunch. I love the ocean here. But for a great vacation, a lake will do just fine. And saying hello to my neighbor isn't a waste of my time.

How does all of this affect my practice as a physician: I grew up in a community that believes that human life is sacred, beginning with conception. This has many consequences for my beliefs and actions as a physician that benefit my patients. It might be helpful to have physicians available with a variety of belief systems so that every patient can find a doctor they can relate to when they make important decisions about their health care. I encountered many situations where this was the case.

I wished that we could get to know each other better: Our assumptions about the lives of the others may be wrong. People we believe to be smart may not be as smart as we think. People labeled as stupid and backward may actually have great ideas. Other communities are not failed attempts of being like us. We can learn from each other.

Student Politician goes Cross-Country

In the summer of 2018, ASSU Senator Matthew Wigler, '19, took a road trip to America's swing districts to learn about the voters who rejected partisanship in a time of great political polarization.

Wigler began his road trip and political research project at Stanford the day before Independence Day. He traveled coast-to-coast in a vehicle he borrowed from a family friend, ending the trip in late August in New York City. In the name of bipartisanship, he traveled with Michael Gofman, executive director of the UC Davis College Republicans, for the first 16 days of the trip. Together they visited gathering spaces like restaurants and coffee shops to interview swing voters and learn about their political views, their lives and concerns.

The goal of Wiglers' research was to understand how voters in these districts have withstood partisanship and kept their fingers on the pulse of both red and blue America at once by remaining open to both sides.

"Learning how the people of 'purple' America think, and how they entertain and act on both Democratic and Republican perspectives, may offer us a model by which we partisans might engage with one another as well," Wigler said. "Ultimately, it's the quest in search of common values that underlies the whole of this enterprise."

The full article by Alex Kekauoha, Stanford News Service, can be accessed here:
https://news.stanford.edu/2018/07/05/student-politician-goes-cross-country/

Do You Discriminate Along Party Lines?

An interdisciplinary research team at Stanford University, Tel Aviv University and the University of Pennsylvania conducted real-world field experiments to understand, whether political partisanship affects the selection of co-workers or employees in an online labor market. They found a strong preference of employers and employees to work for co-partisans. When provided with different economic choices, workers with strong partisan attachments and weaker partisans both demonstrated a preference towards products or services from co-partisans.

Dr. Neil Malhotra, the Edith M. Cornell Professor of Political Economy in the Graduate School of Business and Professor of Political Science (by courtesy) in the School of Humanities and Sciences at Stanford summarized: "There is potential for our findings on partisan-based discrimination to be more troubling than racial- or gen-der-based discrimination because there are no social norms against it," he said. "There's no real social shaming if you don't hire someone because she's a Democrat or Republican, but do we really want a society segregated by partisan identity?

The Economic Consequences of Partisanship in a Polarized Era

By Christopher McConnell, Yotam Margalit, Neil Malhotra, Matthew Levendusky
American Journal of Political Science. January
2018, Vol. 62, Issue 1, Pages 5-18

"Conservatives and liberals can find common ground."

Jesse Jackson

"Life is really simple, but we insist on making it complicated."

Confucius

"You have not converted a man because you have silenced him."

John Morley

"A small group of determined and likeminded people can change the course of history."

Mahatma Ganhdi

DIVERSITY OF THOUGHT

"The Happiness of your Life Depends upon the Quality of your Thoughts."

– Marcus Aurelius

Diversity of Thought

Our final chapter celebrates "diversity of thought", which forms the very center of our diversity initiative. Diversity of thought refers to the concept that each human being has a unique blend of identities, cultures, and experiences that inform how he or she thinks, interprets, negotiates, and accomplishes a task. Diversity of thought enables us to generate a broader spectrum of creative solutions to clinical and scientific problems in the STEM field. Great problem solvers leverage experiences from other fields or disciplines to make conceptual leaps.

The stories in this chapter demonstrate how diversity of thought can influence and enrich the STEM community. The interested reader will learn how experiences from different contexts can inform our work as clinicians and researchers. Diversity of thought can facilitate "out of the box" thinking as an antonym to bias: While bias uses personal judgment to create shortcuts to our thought patters and disconnect people, "out of the box" thinking combines different experiences to produce new insights and expand our collective horizon.

Importantly, our thoughts and perceptions are constantly changing. What might seem impressive today might be replaced by the next innovation tomorrow. I recently read about an interesting experiment: People were asked how much they were willing to pay to see their favorite band from 10 years ago today and how much they were willing to pay to see their current favorite band in 10 years from now. Interestingly, people were willing to pay more to see their current favorite band in 10 years from now than they were to see their favorite band from 10 years ago today. It seems that we underestimate our own evolution.

How to be an ally

An ally is a person who actively promotes and aspires to advance a culture of diversity and inclusion through intentional, positive and conscious efforts. The points below provide examples, how effective allies can promote and nurture diversity of thought, creativity and innovation. These points are not meant to be comprehensive and may require adjustments for different communities and contexts:

- Have an intentional recruitment strategy
- Ensure job security for as many team members as possible. People who are anxious to survive cannot be creative.
- Create an environment where everyone feels safe to share their opinions and ideas.
- Facilitate submission of new ideas by team members through various channels, such as discussion groups, email communications, surveys, anonymous feedback boxes, etc.
- Facilitate collaborations between team members from diverse backgrounds
- Letting one person select all members of a committee can be a major source of bias and/or foster office politics. Introduce an approach to assign committee members in a random fashion, e.g. by a computer algorithm. The algorithm could consider meta-data, such as a minimum number of committee members with a specific characteristic.
- Introduce rotating leadership roles and rotating moderator roles at meetings.
- Actively seek opportunities to lift others up by advocating for them and/or their cause.
- Build trust by understanding the issues faced, acting selflessly, making sacrifices for the group and remaining loyal to the group over time.
- Democratize submission of and access to data within your organization so that team members can develop ideas and make decisions informed by more diverse data sets
- Share growth opportunities with others.
- Provide members from underrepresented minority backgrounds with opportunities to share their experiences and ideas. Listen.
- Avoid controlling meetings or stating your own opinion at the outset of a discussion.
- Offer educational workshops that promote tolerance of multiculturalism.
- Seek opportunities to engage with people who have different opinions and experiences than your own. This will enable you to learn from different points of view and diverse ways of seeing life.
- Ask questions. Be genuinely curious about others and enjoy learning about them.
- Do not burden others with requests to talk about traumatic experiences they haven't shared.
- For communications that may be distributed to people with dyslexia, download dyslexie font and include it in your Word, Excel or PowerPoint software. While easier to process by people with dyslexia, dyslexie font is still quite readable by anyone.
- Make space for other voices and perspectives in meetings.
- Be transparent with your team. Minimize speculations as much as possible.
- Collaborate with other teams, departments and institutions. Foster networking opportunities.
- Actively look for and remove barriers that constrain a colleague's potential to succeed
- Whose books are you reading and whose presentations are you attending? Support creatives from diverse backgrounds by buying their work, attending their presentations, reading their articles/books and contributing to their initiatives and campaigns

The Value of Multidisciplinary Teams: Strength through Differences

As a senior faculty member, a woman, and a member of a minority group, I am both delighted and excited about our department's current activities and efforts in promoting greater inclusiveness and respect for diversity in our work place. At Stanford, we are privileged to work in an academic environment where the value of multidisciplinary teams composed of individuals with differing disciplines, perspectives, and experience, has long been recognized and where "strength through differences" has served as a core tenet that has contributed mightily to our institution's successes.

Each individual (faculty, trainees, university and hospital staff) in the Department provides a unique and important contribution to our collective missions; luckily, each of us comes with a different personal history and cultural identity that has been shaped by a myriad of influences including our age, ethnicity and gender. I believe that our department's ability to engage and fully benefit from the talents and creativity of our constituent members will be commensurate with our willingness to encourage, hear, and act upon their diverse viewpoints.

Thanks to Dr. Sam Gambhir for his commitment and support for inclusiveness and diversity within the Department as well as to Dr. Heike Daldrup-Link for her passion and leadership in heightening our awareness of and sensitivity to these important issues.

Ann Leung, MD

Professor of Radiology
Associate Chair, Clinical Affairs
Division Chief, Thoracic Imaging
Stanford Medicine | Radiology

"The truly creative mind in any field is no mo e than this: A human creature born abnormally, inhumanly sensitive. To him... a touch is a blow, a sound is a noise, a misfortune is a tragedy, a joy is an ecstasy, a friend is a lover, a lover is a god, and failure is death. Add to this cruelly delicate organism the overpowering necessity to create, create, create—so that without the creating of music or poetry or books or buildings or something of meaning, his very breath is cut off from him. He must create, must pour out creation. By some strange, unknown, inward urgency he is not really alive unless he is creating."

Pearl S Buck, American novelist
and recipient of the Pulitzer Prize and the Nobel Prize

Cultural Respect

When discussing diversity, we sometimes fail to fully acknowledge that it also includes respect for other cultures and sensitivity to customs that may appear different to ours and sometimes feel difficult to comprehend.

Growing up with a father who spend his time traveling through the world negotiating deals and founding companies in Europe, Asia, South America and Africa I remember the time he took preparing each of these travels, by carefully studying each country's specific customs. I learned that in China it is considered an insult if food is not left behind at a banquet and that in Japan exchanging business cards is a ritual, which also includes accepting the other person's card with both hands. Also, Brazilians do not speak Spanish but Portuguese and they appreciate pronouncing words accordingly. We should also know that there may be complex relationships between certain cultures such Argentinians and Brazilians as well as Turkish and Greek people.

While some of these issues may appear insignificant at first sight, true diversity requires respect and a willingness to acknowledge cultural differences. We need to be willing to accept other people, their unique culture and their customs. Diversity is not about promoting our perceived interpretation of how we feel diversity has to look like or using it as an instrument to gain political power. It is intimately intertwined with respect, tolerance and honesty and true diversity will not exist without these values.

Thomas M. Link, M.D., Ph.D.

Professor and Division Chief
Musculoskeletal Radiology
Department of Radiology and Biomedical Imaging
University of California, San Francisco

Human Spirit from Around the World

Dr. Martin Willemink, instructor at Stanford Radiology, expresses his creativity in photography. Here are examples of his photos, which he took in different parts of the world and which show the beauty of diversity.

Let's reboot our curiosity!

Imagine you could have all the information ever known to humankind, instantly available, at any place and any time. This is our reality today. Whether we want it or not, we are flooded with information every day, through hun-dreds of notifications, emails, phone calls, text messages, social media, and chat boxes on our PACS workstations. We are hyper-saturated with unsolicited information and we invest a huge amount of energy sorting through irrel-evant information and unwanted noise.

The bitter consequence is this: There has never been a time where we could explore more and learn more. But since we are overwhelmed with information influx, we have less and less time to think and actively seek new dis-coveries. By constantly processing incoming information, we kill our curiosity. We cannot possibly go out there and seek more. If we add one more piece of information, our brain might explode. There has never been a time where we were engaging more and imagining less. We are getting caught up in the past by constantly sorting through yes-terday's information.

Curiosity, imagination and creativity can solve problems, improve situations, build relationships and discover new and better worlds. Let's focus on tomorrow. Let's re-boot our imagination and rebuild our gleaming curiosity!

Anonymous

The Creative Spark at Stanford

Creativity is a process that involves the identification of hidden patterns followed by the connection of seemingly unrelated ideas in unique and exciting ways to produce tangible results. It is often refined by the application of knowledge, divergent thinking skills and exposure to new experiences. My creativity is sparked by the environment with generous doses of nature and art. There are plenty of opportunities to experience this at the Stanford University campus. From the serene Sculpture Gardens, to the panoramic view of the Bay from the Hoover tower and the captivating pieces in the Cantor Arts Center that transport you to various regions in Africa, Europe and Asia. All of which nurture my creativity and allow me to approach research from a different perspective. Frequently, scientific research is viewed as a linear and logic process to solve problems. We identify an unmet need, propose a hypothesis, design experiments to test our hypothesis and acquire data that is presented in a structured way. While this rigorous approach is valued, stepping away from the normal routine in the early stages may create an opportunity to explore the research problem from different viewpoints to consider alternative solutions.

The creativity process in research can be stifled by the limited availability of funding to explore new ideas and produce tangible results. This limitation is a major concern for early career researchers. However, the Radiology Department is actively involved in supporting researchers by allocating vital funds for the advancement of science. It is encouraging to know that the Radiology Department has measures in place to fuel creative sparks.

Louise Kiru, PhD

Instructor
Stanford Medicine | Radiology

Magic meets Science

Would you be shocked if I were to tell you that the scientific community has a tremendous amount to learn from the world's magicians about how to do better science? These two communities might seem worlds apart, with one focused on truth and discovery while the other is focused on secrets and deception. But dig a little deeper and you will find that magicians and scientists have a lot in common—we both attempt to push the boundaries of what's possible! For example, one of the most miraculous examples of brilliant science cultivating the impossible can be found in Seaborg's classic paper, "The Energy Dependence of Bi Fragmentation in Relativistic Nuclear Collisions." Charlatan alchemists attempted for millennia to turn lead into gold - Seaborg made it happen.

Magicians have a few specific approaches for cultivating the impossible. One approach begins with a vision for an impossible outcome (e.g. making an assistant levitate, or an object disappear). Starting with this vision, the magician then spends years developing methods to make the impossible a reality. A second approach begins with a method (i.e. a way to move an object from one place to another), followed by exploring applications of that method. These approaches that magicians take to realizing the impossible are not so different from what we do at the Canary Center, where we work diligently to find methods to make cancer disappear. By studying the way that magicians approach a new illusion, the scientific community can likely find new ways to approaching science.

Beyond dreaming of the impossible—magicians are among the world's foremost students of perception, constantly experimenting on their cohorts (audiences) to examine how biases and perceptual gaps lead our brains to perceive miraculous events that never actually occurred. Teller, of the famous duo Penn and Teller, said it well: "At the core of every trick is a cold, cognitive experiment in perception."

Magicians know that their hands can never be faster than your eyes. However, they also know that the brain is a well-intentioned liar. These lies are essential to our everyday function. Think about your drive to work this morning. At any point did you stop at a stop sign? Your brain took in some observational (optical) information and then used that information to infer the presence of a stop sign, triggering you to stop. What would have happened if you were only able to see half of the stop sign, either because it was blocked by another car or a tree branch? Would you have still stopped? More than likely, yes. Despite having incomplete information, your brain is drawing on its experience and making a series of assumptions, filling in the blanks and perceiving a stop sign—even when you haven't actually seen the entirety of one.

In the real world, the distinction between reality, observation and perception is often fuzzy and under-appreciated. Magicians exploit this distinction to amuse and delight, by inducing you to perceive that the impossible actually occurred. Scientists must be acutely aware of this distinction to ensure the quality of their science. Every time we as scientists attempt to look at data "objectively," our brains are filling in the blanks, over-interpreting and bringing our biases to bear. The same factors of expectation-bias and change-blindness that are the foundation of many magic tricks also ruin many scientific studies.

Take it from a magician-scientist—we scientists have much to learn from the world's magicians.

Parag Mallick, PhD

Associate Professor
Stanford Medicine | Radiology

Be here now

"The best and most beautiful things in the world cannot be seen or even touched. They must be felt with the heart."

— *Helen Keller*

Joy is Created by People, Experiences and Identity within a Community

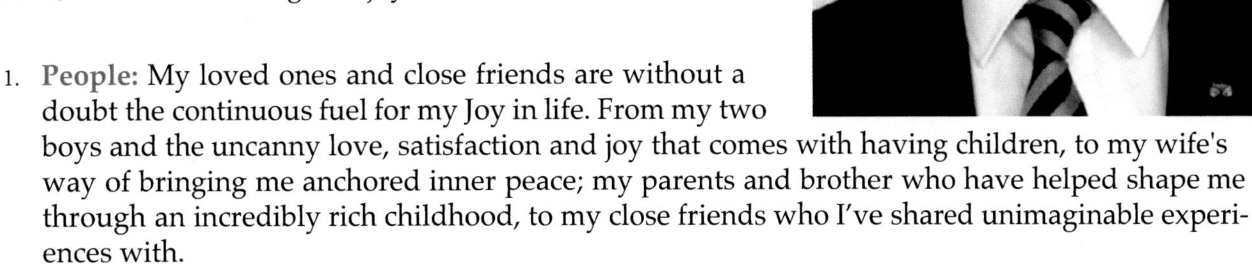

I've never really thought through what brings me joy, or what joy even means to me. The first thing I did when I agreed to contribute my thoughts on this topic was to ask 'Google' what joy is, and how it differs from happiness (I often use both interchangeably). I learned that unlike happiness' reliance on external forces and circumstances (and it's role as a socio-economic metric), joy is a permanent inner feeling that defies circumstances (and difficult to measure).

Living with joy is often consistent and long lasting, and it results from our inner peace and satisfaction. To me, joy is a continuous manifestation of the people, experiences and identities that define who we are and help 'ground' us. Based on this, here is what brings me joy in life:

1. **People:** My loved ones and close friends are without a doubt the continuous fuel for my Joy in life. From my two boys and the uncanny love, satisfaction and joy that comes with having children, to my wife's way of bringing me anchored inner peace; my parents and brother who have helped shape me through an incredibly rich childhood, to my close friends who I've shared unimaginable experiences with.

2. **Experiences:** From unpredictable itinerary-less backpacking travel packed with adventures in strange lands, and culture shocks that i continue to be grateful for, to playing live music in front of large audiences as a teenager. And while experiences are not all created equal, together, they make up the pieces of the bitter-sweet journey to recognize joy.

3. **Identities:** We all carry in us a variety of identities. From cultural origins, spiritual beliefs or lack of, sexual identity, to the complex communities that we belong to, either professionally or not. I find joy in the multiple identities that make up the blueprints of who I am, and in getting to know those of others. For me, it's the french-canadian, egyp-tian-origin, engineer-scientist, entrepreneur, with a passion for the philosophy of science, spirituality, music, and exploring the unknown.

As such, I consider myself fortunate that I can continue to build on those elements of joy within the Department of Radiology at Stanford, surrounded by incredible colleagues, friends, with rich diversities and identities, who are eager to connect and share experiences that are personally and professionally impactful.

Ahmed El Kaffas, PhD

Instructor
Stanford Medicine | Radiology

The Power of Mental Habits

As we reflect on creativity, joy and happiness in this chapter, I know that we will be yet again reminded of the power of our mental habits which only we can cultivate, thought-by-positive-thought. Rather than being something that we achieve, happiness is something that we cultivate and maintain (like Sisyphus.) One of the many ironies of life is that: this is more difficult to do when we need it most. I'll focus this contribution on the six 2 minute daily practices of Shawn Achor. They are not unique, but they are concise and easy (aka a low bar.)

Gratitude

The first and foremost is gratitude. The chances are good that if you are at Stanford and therefore reading this, like me, you may be living a life of extreme privilege relative to most people on the planet. Unfortunately, it does not follow that we live in a state of constant extreme gratitude or happiness. In a 2017 study, while a rise in household income from $40 to 50k correlated with an increase in happiness, increases above $75k did not. However, simply identifying 3 things from the previous 24hrs for which we are grateful does measurably increase our happiness, so here goes:

- Talking with my octogenarian friend, Ellie (85%)
- Witnessing the generous spirit of my husband, Joe (90%)
- Taking deep breaths of clean fresh air (75%)

This may not be for everybody, but I also "quantify" how grateful I feel for each item (on a completely arbitrary and subjective scale) and for some reason that makes the list more impactful for me.

The Doubler

Another daily practice that measurably increases happiness is journaling for 2 minutes about a positive experience. Here's mine:

One night during the shutdown, I "slept" with my 9yo daughter, indulging her request, and one of the many times that I awoke, she was laughing in her sleep. I stayed awake watching and listening to her in fascination and admiration, and giggling quietly to myself. She loved hearing about it the next morning and we got to laugh about it again together (this time, both of us were awake.)
For some reason, writing about the experience is more effective than either telling somebody about it or re-membering it.

Cardio

If you have the time (wink wink), doing a whopping 15 minutes of fun cardio activity daily also increases our subjective experience of wellbeing. These days, it is especially important to remember to move. It's scary how easy it is not to step foot outside the house (possibly because that requires getting out of pajamas.) I like to take a walk while talking with a friend on the phone. This gives privacy for the call as well as the cardio-hap-piness benefit and deepens social connections (see below.) Others like to listen to music, a podcast or audio-book, or even enjoy their immediate surroundings with nothing in their ears. If you prefer to stay in your paja-mas, the occasional dance party, either alone or with the kids, is a potent (and potentially comedic) happiness booster. 15 minutes of yoga (I do 5 sun salutations) in the morning is another pajama-friendly cardio activity, and is like moving meditation.

Meditate

As little as 2 minutes of meditation a day produces a measurable increase in happiness if practiced consis-tently. This is not to be confused with the time it takes to remember what you wanted to get from another room once you've gotten there, or simply spacing out. Those don't count. As you have heard a gazillion times, focusing on your breath is a good way to start meditating.

Kindness

Another sure-fire daily exercise to strengthen your sense of wellbeing is to send someone you know a short email or text appreciating them. If it's the kind of day when nobody seems praiseworthy, take food to a friend in need or offer to shop for an older neighbor. Keep in mind that this will probably benefit you more than them (unless you save them from contracting COVID-19).

Relationships

If, in addition to being happy, you'd like to live a successful, healthy, and long life, invest in meaningful social connections. This is why humanity has evolved with interactions through churches, sports (and their various forms of worship), self-help groups, volunteer organizations and academic communities. Granted, zoomx-haustion is real. But the benefit of human connection outweighs the computer headache. It took me a while to warm up to virtual connections with people but on the upside you can stay in your pajama bottoms.

Challenge often brings gifts that are not normally accessible. 40-60% of our tendency to feel happy may be genetic. While my genetically high metabolic rate lets me get away with eating more calorie-rich foods than many can, my disposition does not allow for analogous mental delinquencies. If your family has a history of depression or suicide, learning to function with that emotional burden builds a strength that people with hap-py-genes may never have the opportunity to develop. Physical isolation with our thoughts (and possibly our family) may occasionally bring us to our knees during social distancing. With this challenge, we may be more receptive to investing in positive habits than we are under normal circumstances. This unusual context may also make us more aware of the benefits of these positive practices. While sustained perfect happiness does not exist, we can enjoy progress in that direction with a few simple practices. So, now is the time: write that email, make that call, crank up the music... Breathe.

Laura Jean Pisani, PhD

Associate Director, SCi^3
Stanford Medicine | Radiology

*"Joy lies in the fight, in the attempt,
in the suffering involved, not in the victory itself."*

– Mahatma Gandhi

"The person born with a talent they are meant to use will find their greatest happiness in using it."

— Johann Wolfgang von Goethe

Stanford Radiology Diversity Fair

The Radiology Department hosts an annual diversity fair, featuring food and cultural traditions from different continents along with a world map, where attendants can mark their home town. The event is kicked off by a Grand Rounds lecture. At the event, representatives from the Office for Faculty Diversity & Development at Stanford Medicine and the Provost Office provide information about diversity resources at Stanford. At our most recent diversity fair, more than 200 people from the Department of Radiology and many other Departments attended.

Closing Statement:

Uniting Diverse Minds Inspires Innovation

Our goals build on Stanford's diversity charge : "Many of our most important learnings come from hearing and respecting diverse perspectives. Equity is a fundamental goal as we seek together to create a society of fairness, character, and integrity."

Humans filter and judge all incoming information. Jo Brand pointed out that everyone who has had the opportunity to read a press article about themselves knows that there is a huge amount of inaccuracy, judgment and editorial bias that reflects the views of the writer rather than those of the subject. Most of us do not question the status quo but are critical of the unknown. In addition, while the majority of us hold ourselves accountable to a high standard of integrity, there are outliers and trolls in every community including ours. This is not only ethically unacceptable but also poses a risk for our organization as information about discriminatory or otherwise unacceptable behavior can rapidly spread through social media channels and damage our collective reputation. Winston Churchill pointed out: "The truth is incontrovertible. Malice may attack it, ignorance may deride it, but in the end, there it is." We seek to take preventive action by identifying and addressing potential areas of bias and creating a supportive platform where concerns can be processed within the department.

To make our workplace supportive, just and inclusive for everyone, we are creating leadership teams with a broad representation of qualified people from diverse backgrounds in order to ensure that every member of the Department has an advocate at the leadership table. Michael Spencer said: "Sexual harassment is a symptom, one dimensional leadership is the cause. A lack of inclusion, diversity and gender-balance in leadership has direct impacts on the quality of the culture for women and other minorities". Introducing powerful advocates for everyone at the leadership table reduces the risk of discrimination and harassment at the workplace and provides our students and junior faculty with a diverse set of role models. Considering the increasing complexity of diversity in terms of race/ethnicity, economic background, age, gender, sex and sexuality, among others, we seek to create a fundamental framework that allows us to understand and continuously adjust our faculty ensemble. We aim to create opportunities for every team member to learn, grow and advance in a supportive environment.

We ultimately seek to unite diverse minds to inspire medical innovation: Our experiences, abilities and perspectives are unique. Nobody can use them as we can. We should be asking our leaders what they will do with their team members, not what they think of them. We are rewiring our minds from talking about people to talking with people about great ideas. At Stanford, we are blessed with a concentrated assembly of gifted minds who have access to abundant resources and opportunities. Millions of people would like to be in our position. Do we live up to their expectations? Do we use our combined intellectual capacity and resources to make their life better? I hope that insights from this book have inspired the interested reader to enable diverse teams to create progress and make an impact on the lives of patients around the world.

H. E. Daldrup-Link

Heike E. Daldrup-Link, MD
Professor and Associate Chair for Diversity
Stanford Medicine | Radiology

"We delight in the beauty of the butterfly, but rarely admit the changes it has gone through to achieve that beauty."

– Maya Angelou

Thank You!

We thank all of our contributors for investing the time and effort to share their experiences and thoughts with us. Your insights enrich our Department and the broader STEM community!

We also thank members of our Radiology diversity committee as well as many colleagues, trainees and staff, who critically reviewed this book and provided important feedback and comments, that helped to improve its design and content. We specifically thank the following individuals for critically reviewing the final version of the book:

David Larson
Benedict Anchang
Guido Davidzon
Terry Desser
Ann Leung
Sandy Napel
Jayne Seekins
Tanya Stoyanova
Bonnie Maldonado
Susan Kopiwoda
Will Kwong
Jamie Tran
Mekemeke Faaoso
Jessica Klockow
Aisling Chaney
Wendy DeMartini
Jim Strommer
Abraham Verghese

We thank Amy Thomas for the design of this book. Her unique approach to combining written communications and visual art helped our authors to effectively convey their messages. The beautiful imagery in our book has been generously provided by the individual contributors below, as well as, by the following sources:

Mark Tuschman
Kensley Villavasso
Berthold Schroeter
Larry Chow
Martin Willemink
Michael Federle
unsplash.com
pexels.com